Oxford's Savilian Professors of Astronomy

Sir Henry Savile's monument in the antechapel of Merton College shows him flanked by Ptolemy and Euclid, the two foundational authorities whose work he expounded in his Oxford lectures.

OXFORD'S SAVILIAN PROFESSORS OF ASTRONOMY

The First 400 years

EDITED BY

ROBIN WILSON

STEVEN BALBUS

Great Clarendon Street, Oxford, OX2 6DP,
United Kingdom

Oxford University Press is a department of the University of Oxford.
It furthers the University's objective of excellence in research, scholarship,
and education by publishing worldwide. Oxford is a registered trade mark of
Oxford University Press in the UK and in certain other countries.

Published in the United States of America by Oxford University Press
198 Madison Avenue, New York, NY 10016, United States of America

British Library Cataloguing in Publication Data
Data available

Library of Congress Control Number: 2025940423

ISBN 9780198894292

DOI: 10.1093/oso/9780198894292.001.0001

Printed and bound by
CPI Group (UK) Ltd., Croydon, CR0 4YY

The manufacturer's authorised representative in the EU for product safety is
Oxford University Press España S.A. of Parque Empresarial San Fernando de Henares,
Avenida de Castilla, 2 – 28830 Madrid (www.oup.es/en or product.safety@oup.com).
OUP España S.A. also acts as importer into Spain of products made by the manufacturer.

FOREWORD

The Savilian professorship of astronomy, the oldest university chair of astronomy in the world, was established in 1619. Its founder, Henry Savile, despaired of the quality of mathematics education in England at the time, not entirely without cause. College tutors were generally trained in the classics and often possessed little more than a rudimentary familiarity with the 'mathematical arts'. Savile took it upon himself to rectify this lamentable state of affairs by endowing the two positions that to this day bear his name: the Savilian professors of geometry and astronomy.

In the 17th century, astronomy was considered a very mathematical subject, much more so than it is today. It is not difficult to see why. At the time, the subject was overwhelmingly concerned with ascertaining accurate positions of celestial objects, and any readers who try to calculate the angle at which the sun sets relative to the horizon in Greenwich on (say) 23 November will quickly grasp the difficulty of coping with a web of spherical trigonometric manipulations. For the first 250 years of the astronomy chair, its holders tended to be exceptionally skilled mathematicians, two of whom (Abraham Robertson and Stephen Rigaud) held both the astronomy and geometry chairs. Herbert Hall Turner, a polymath who both literally and figuratively brought the chair into the 20th century, was notable for the breadth of his scientific contributions, from innovative astronomical data reduction techniques to geophysical seismology.

While Christopher Wren is probably the best known of the Savilian professors of astronomy, the most distinguished holder by virtue of his contributions to the field is certainly James Bradley. Bradley's name is familiar to many students of physics for his explanation of the aberration of starlight, a consequence of the Earth's orbital velocity. Once understood, by measuring the induced shift in stellar positions (~20 arc-seconds), the speed of light could be determined to an accuracy of 1 per cent. Before Bradley's work, it was known only to within a factor of 2 or so.

Bradley's superb observational techniques allowed him to measure the Earth's nutation (a very small wobble associated with the precession of the equinox) for the first time. Less well known, but of at least equal importance, was his ability to control and reckon quantitatively with observational errors, a task that today would be done with sophisticated statistical techniques. This ultimately made possible the great 19th-century achievement by Friedrich Bessel (another mathematician–astronomer) of the first stellar parallax angle determinations, measuring the distances to nearby stars by triangularization techniques, using the Earth's orbital diameter as the baseline. To this day, these measurements are foundational to all astronomical distance determinations.

The 20th century witnessed the discipline of astronomy establishing its own niche within physics, with the birth of astrophysics. This initially focused upon stellar spectroscopy, which allowed elemental abundances of the Universe to be reliably determined; with Harry Plaskett and Donald Blackwell, the chair had two prominent practitioners of this important discipline. But by the 1980s astrophysics had exploded, literally so with regard to the Big Bang theory of the Universe becoming firmly established. George Efstathiou and Joe Silk fit seamlessly with these times, bringing Oxford Astrophysics to the front lines of cosmological research. When I assumed the chair in 2012, I was at once daunted and pleased to have the chance to be part of this ongoing story. My interests are broad, from black holes and stellar interiors to ocean tides and evolution. When not long after my arrival I found myself giving a colloquium in the Department of Zoology on the last of these topics, I couldn't help but wonder whether Henry Savile would tell me to stop messing about and get back to trigonometry. Having now retired, however, I am confident that the Savilian chair will continue to be a vigorous and influential part of astronomy, long after my time.

Steven A. Balbus FRS
Savilian Professor of Astronomy, 2012–24.

CONTENTS

THE SAVILIAN PROFESSORS OF ASTRONOMY

1619 **John Bainbridge** (1582–1643)
1643 **John Greaves** (1602–52)
1649 **Seth Ward** (1617–89)
1661 **Christopher Wren** (1632–1723)
1673 **Edward Barnard** (1638–96)
1691 **David Gregory** (1659–1708)
1708 **John Caswell** (1656–1712)
1712 **John Keill** (1671–1721)
1721 **James Bradley** (1692–1762)
1763 **Thomas Hornsby** (1733–1810)
1810 **Abraham Robertson** (1751–1826)
1827 **Stephen Peter Rigaud** (1774–1839)
1839 **George Henry Sacheverell Johnson** (1808–81)
1842 **William Fishburn Donkin** (1814–69)
1870 **Charles Pritchard** (1808–93)
1893 **Herbert Hall Turner** (1861–1930)
1932 **Harry Hemley Plaskett** (1893–1980)
1960 **Donald Eustace Blackwell** (1921–2010)
1988 **George Petros Efstathiou** (b. 1955)
1999 **Joseph Ivor Silk** (b. 1942)
2012 **Steven Andrew Balbus** (b. 1953)

The elaborate memorial in the antechapel of Merton College to John Bainbridge, the first Savilian professor of astronomy.

CHAPTER 1

Sir Henry Savile and the early professors

ROBERT GOULDING

This chapter addresses the state of mathematical instruction in Oxford, both before and after the founding of Sir Henry Savile's chairs in astronomy and geometry in 1619. Discussing the terms and conditions of Savile's foundations, we set Savile's benefaction within the general context of Oxford teaching and learning structures. Savile personally appointed his first two professors, John Bainbridge (astronomy) and Henry Briggs (geometry), and we describe their performance and publications, and those of their Savilian successors, John Greaves and Peter Turner.

Astronomy in 16th-century Oxford

It is difficult to pin down exactly the extent to which the mathematical sciences were practised at the University of Oxford in the 16th century. Over the seven years that students would ideally be at the University (taking four years for their BA degree and three years for their MA), they were expected to attend a sufficient number of lectures in the seven liberal arts prior to receiving their bachelor's degree, and to spend their master's degree in the study of the three philosophies (moral philosophy, natural philosophy, and metaphysics). In practice, the distribution of the subjects between the two degrees varied constantly in the University statutes.

Henry Savile matriculated in 1560, at the age of 11 (not unusually young for the period). The statutes that came into effect at the end of his bachelor's degree required the undergraduate students to attend lectures on grammar, rhetoric, and dialectic (the 'trivium') for a total of eleven terms of study. To complete their BA degree, the students needed three terms of arithmetic and two of music, and then, continuing on to the MA degree (the default expectation), they would take two terms of geometry and two of astronomy, thereby completing their study of the four mathematical sciences of the 'quadrivium'. They rounded out the rest of their master's degree with eight terms in the three philosophies.[1] On graduating, the students would then be

Robert Goulding, *Sir Henry Savile and the early professors*. In: *Oxford's Savilian Professors of Astronomy*. Edited by: Robin Wilson and Steven Balbus, Oxford University Press. © Oxford University Press (2025). DOI: 10.1093/oso/9780198894292.003.0001

Two pictures of Henry Savile. (*Left*) As a younger man: according to John Aubrey, he was 'an extraordinary handsome and beautiful man, no lady had a finer complexion'. (*Right*) As an older man, from an original picture in the Bodleian Library.

expected to remain at the University for two years as a 'regent master', delivering these required 'ordinary' lectures to the next generation of undergraduate students.

In his own ordinary lectures on astronomy of 1570 (described below), Henry Savile confirmed that he expected that the students who turned up to hear him lecture on Ptolemy's *Almagest* would have already studied three terms of geometry and two of arithmetic. We have little evidence of what was taught in these classes. Students who came to study at Oxford, even if not quite as young as Savile, would have had no formal training in mathematics from their grammar schools. Most of them would be unable to do simple arithmetical calculations, while some, as Savile noted in his lectures, may not even be able to count. We should not expect that the lectures in the mathematical sciences reached a high level, and indeed, they seem to have been taught in a perfunctory way out of very elementary textbooks.

Most students who reached the lectures on astronomy, the final stage of the quadrivium, would expect to hear an exposition of Sacrobosco's *De sphaera* (On the Sphere) – that is, if they reached that point at all. Through the course of the 16th century, various pressures were exerted on the study of the quadrivium within the seven-year BA/MA programme. Increasingly many students were attending Oxford with the intention of receiving only the BA degree. Wealthier students, in particular, did not wish to commit to the MA degree as well, especially if that entailed remaining at the University to teach. Some left without a degree after only a year or two, having made the all-important social and political connections they needed during that short time. If they attended classes, these students were far more interested in the study of the linguistic arts of the trivium (which now occupied the first years of the degree) than the

'Cista mathematica', Henry Savile's triple-locked chest for housing his mathematical instruments, was part of his provision for the study of the mathematical sciences at Oxford.

quadrivium. Those who continued to the end of the bachelor's degree frequently obtained dispensations from attending mathematical lectures and, as fewer students required these lectures and fewer MAs were prepared to stay on to teach, the administration of these lectures fell onto the shoulders of a small number of newly minted MAs (including Savile).

Thus, seen from one perspective, the system of formal instruction of the mathematical sciences in 16th-century Oxford was in a process of decline, something that Savile himself affirmed several times in his lectures and eventually would remedy by establishing his two professorships. But we may also view this as a shift of location and form of mathematical instruction, for while the formal teaching of the quadrivium in the University schools seemed to be increasingly irrelevant to students and poorly taught, the colleges were becoming the sites of richer and more informal teaching and study of these subjects.[2] Much is still unknown about the kinds of work carried out in the colleges, but it is clear enough that Savile would probably have found some kindred spirits outside the system of official instruction. It is also significant that when Savile founded his lectureships, which were intended to replace the old system of ordinary lectures, he made sure to associate them as well with a particular college, such as his own (Merton College) where they were originally hosted before moving to New College.

Savile's own education was largely solitary – that at least is how he described it in autobiographical passages in his 1570 lectures, though we might expect some exaggeration. It was, in any case, founded on the deep study of ancient Greek mathematics that seemed not to be part

Thomas Allen (1542–1632) of Gloucester Hall (later refounded as Worcester College) was a collector of mathematical books and manuscripts.

of the official University instruction. He had probably attended ordinary lectures on arithmetic and geometry as a BA student, and during his MA degree he would have attended those on astronomy. But going by his own criticisms of them scattered through his own lectures – their reliance on elementary sources like Sacrobosco's *Sphere*, and the inability of many of his fellow students even to add and subtract – he was not impressed. In the Michaelmas term of 1568, Savile took on the mammoth task of translating the whole of Ptolemy's *Almagest* into Latin, together with the commentaries of Theon and (where these were missing) of the Byzantine author Nicolas Cabasilas.[3]

At around the same time, or perhaps a little earlier, Henry Savile became absorbed in the study of Euclid's *Elements*; like most of his contemporaries, he would have encountered this in an edition of fifteen books, rather than the thirteen we meet today. As he completed his *Almagest* in the winter of 1568, he copied into the blank pages of his notebooks the notes he had made on the history of mathematics in the margins of Proclus's *Commentary* on Euclid's Book I, a work appended to the first printed edition of the Greek text of the *Elements*.[4] In his 1570 lectures, he recalled the fresh memory of those years with an unusual gloss. Prompted by an unnamed 'friend', he had thrown himself into the study of Euclid, but had found himself stuck in Book 10 before he could begin solid geometry. Then, according to this account, he devoted himself to Ptolemy's *Almagest* – but he also ran into difficulty there. He then completed his study of the *Elements* and (at the prompting of 'certain men') the works of Archimedes, before returning to the *Almagest*, now with greater success.

Savile's account was based closely on the discussion of the ideal education of the philosopher in Book VII of Plato's *Republic*. There, too, Socrates's interlocutors make the mistake of taking on astronomy too early, before Socrates reminds them that a student should first master the

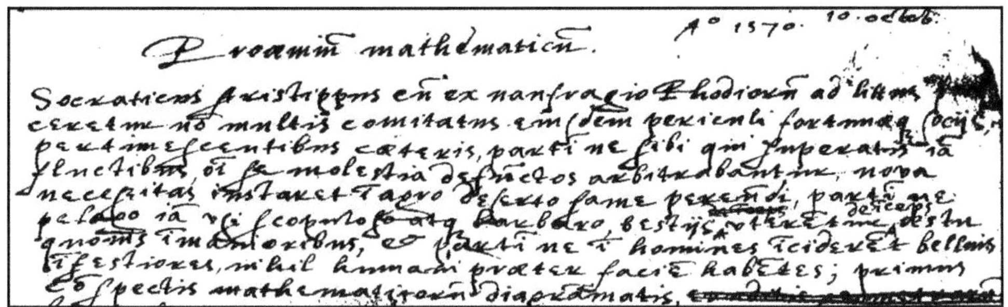

Part of Henry Savile's manuscript of *Prooemium mathematicum*.

mathematics of solid bodies before attempting astronomy, the study of solid bodies in motion.[5] In his account of his mathematical formation, then, Savile imagined himself as the perfect Platonic astronomer, and one who had achieved these heights largely through the solitary study of books, transcending the poor state of Oxford's official instruction.

Even so, Savile acknowledged the encouragement and guidance of others. Who were they? It would not be improbable to include Thomas Allen among them. A few years older than Savile, Allen would go on to encourage many young mathematicians at Oxford over his long career, including Thomas Harriot. Perhaps as a mark of their early mathematical friendship, Savile later appointed him as a Fellow of Eton College.[6]

Henry Savile as an astronomer

Savile established an early reputation through his 1570 lectures on Ptolemy. In their original plan, these lectures began with a *Prooemium mathematicum*, praising mathematics and each of its constituent arts in high Platonic style. These first three lectures established the tone that he wished to maintain throughout his *Almagest* lectures over the coming year: serious, erudite, and far more concerned with fundamental theory than with practical observation. At the end of this third lecture, Savile had promised to embark upon Ptolemy on the next day, but in an apparent change of plan, he mined the notebooks he had kept from the previous couple of years to set out a history of mathematics from Adam to Ptolemy. In the course of five dense lectures, he introduced his students to 154 mathematicians and astronomers.

Finally he turned to the *Almagest*. Moving rapidly through the first cosmographical book, Savile then slowed down to lead his students through basic methods of mathematical calculation. As he progressed through the *Almagest* in 37 lectures over the remainder of the Michaelmas term, Savile set out the fundamentals of the mathematics of the sphere, moving back and forth between Ptolemy's text and more recent works on trigonometry by Regiomontanus, Oronce Fine, and others – and even making a nod towards Copernicus.[7]

In his subsequent notebooks Savile covered the remainder of the *Almagest* and, as he moved to the books on planetary astronomy, he set out a close study of the models of Copernicus's *De revolutionibus*, comparing the new heliocentric models parameter by parameter with Ptolemy's. If he actually delivered these lectures, they would have been the earliest presentation that we know of on the 'new astronomy' at an English university, a quarter-century after its publication.

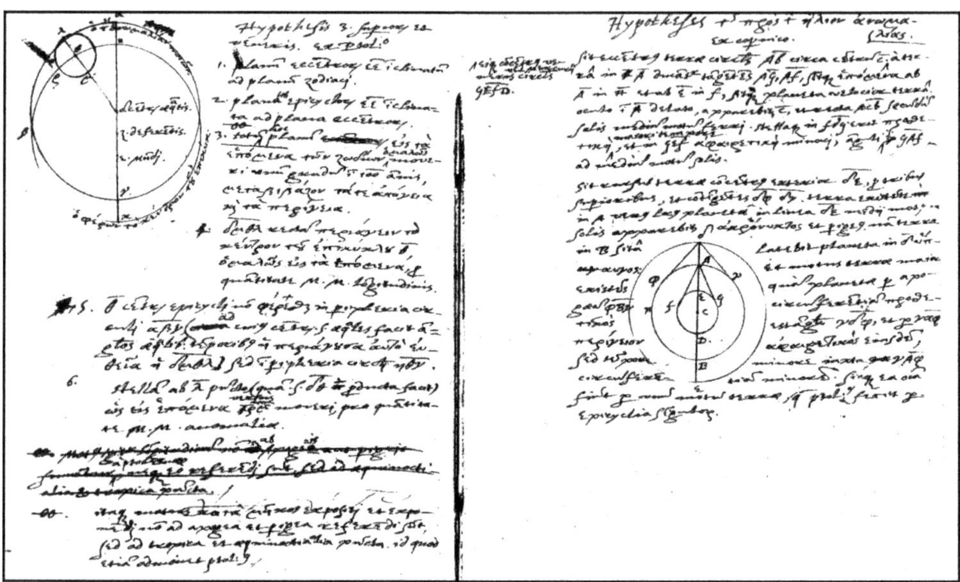

Savile's lectures, the earliest intended public teaching in Oxford of the new astronomy for which we have evidence, presented both the Ptolemaic (*left*) and the Copernican (*right*) models for the Sun's apparent motion.

In fact, Savile's lectures did not resume after the Christmas break, because of an outbreak of plague. The students left Oxford, and Savile and the other ordinary lecturers were excused from all duties for the remainder of the year. In the following year, the University required the lecturers to return to work, but Savile may have received a special dispensation from Merton College that excused him from that year also.[8]

With all of these interruptions, it is unclear whether Savile lectured for more than a single term, and thus his famous lectures may have avoided most of the rigours of technical astronomy and may also have omitted the introduction to Copernicus that he had intended. In their truncated form they were erudite, grounded in humanistic scholarship, and concerned with the elementary foundations of a mathematical and astronomical education. Somewhat ironically, the fortuitous form of Savile's lectures would set the tone for the early Savilian professorship a half-century later. His early professors certainly did not shy away from difficult texts (including the *Almagest*), but their focus would remain largely on the scholarly task of recovering and commenting on ancient texts.

Savile completed his mathematical and astronomical education in the late 1570s and early 1580s, in a long Continental tour with a few close friends, including Philip Sidney.[9] He spent time in Paris with François Viète (who failed to impress him), and with the astronomers Johannes Praetorius in Altdorf (near Nuremberg) and Tadeáš Hájek in Prague, both of whom were deeply engaged in work on the Great Comet of 1577. He spent the longest time in Wrocław, at the house of the Hungarian Unitarian (and former Catholic bishop) András Dudics, where he worked for some weeks on semi-Copernican planetary models alongside Tycho Brahe's former collaborator, Paul Wittich.[10] His last major stops were in Padua where he stayed at the great library of Gian Vincenzo Pinelli (which later became a favourite haunt of Galileo's), and in Venice where he amassed manuscripts and printed books to bring back to England, and eventually to the Savilian professors' library. This period marked a burst of mathematical activity on Savile's part. As well as his collaborations with European astronomers, he delved deeply into the foundations of the Eudoxan theory of proportion, as it was laid out in Book 5 of the *Elements*.[11]

On his return to England, he found himself increasingly engaged in theological projects, culminating in 1610–12 in his vast edition of St Chrysostom.[12] Even more, however, he found a new interest in the entangled worlds of politics and academic administration. Through his powerful connections, Savile had been made Warden of Merton College in 1585, and additionally Provost of Eton a decade later, ruling both with an iron hand although frequently *in absentia*. While he occasionally interested himself in happenings in the world of mathematics, such as in Scaliger's much-derided attempt to square the circle,[13] he seemed to involve himself in very little actual mathematical work during his middle years. His next great contribution would be the foundation of his two professorships, an act which, in a sense, tried to revive in form and content the mathematics of his active youth and the excitement that he had taken in it.

Ramus and the Ramus chair

In his early years of self-education, Savile had relied primarily on Petrus Ramus's influential *Prooemium mathematicum* of 1567 as his guide to mathematics and its history. Ramus presented a comprehensive history of mathematics, beginning with the biblical patriarchs and lingering in Greek and Roman antiquity – this was the most important single source for Savile's long historical excursus in his Ptolemy lectures. He then went on to describe the recent condition of the mathematical sciences and arts in Europe, extolling the state of the disciplines in German lands. France, and the University of Paris in particular, was in a less happy state – Ramus's *Prooemium* was actually written in response to the effective suspension of the mathematical chairs in the Collège Royal. As for England, Ramus had written to John Dee in 1565, telling him that he was planning to write an introduction to mathematics, and asking him to comment on the state of mathematics in his own country, but while Ramus knew of the two great universities, and even the names of their colleges, he could find nothing about the study of mathematics there, nor about any great mathematical authors or professors at either institution.[14] Dee must have responded negatively for, two years later, Ramus lamented that England was, as ever, the mere pupil of France – even France in its current unenviable state. Addressing Queen Elizabeth, Ramus urged her to establish 'Regius Professors of mathematics in both Cambridge and Oxford, to adorn your memory with eternal praise for your magnificent generosity'.[15]

In his earlier years the young Savile had not been inclined to agree with Ramus, however much his historical lectures had been drawn from Ramus's *Prooemium*, but in his lectures on the *Almagest* he concurred. Drawing his students' attention to the 'recently published criticism' of the University, Savile observed that the teaching of mathematics at Oxford had greatly declined since the heyday of the 'Merton School' (as it is now known), some two centuries earlier. Teachers now were lazy and taught the easiest texts, students cared nothing for the subjects, and the institution itself did nothing to sustain its teaching. He hoped, he said, that he could be the one to restore mathematics at the University, even as a mere ordinary lecturer.[16]

Two years later, Ramus was killed in the St Bartholomew's Day massacre in Paris, and his status as a Protestant martyr instantly raised his profile in England. Like many of his countrymen,

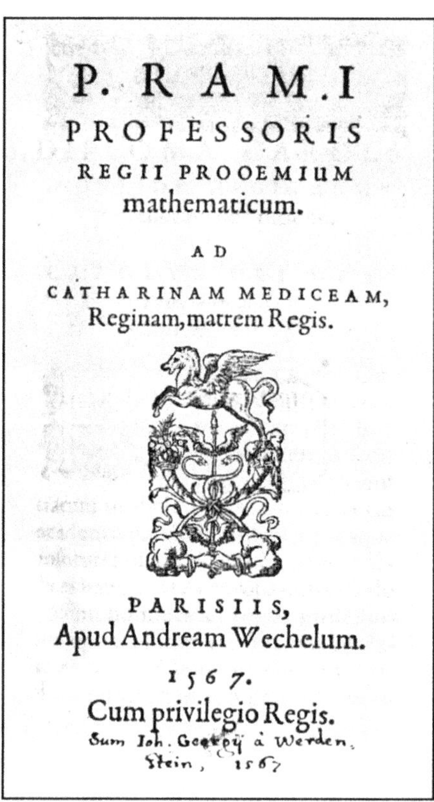

P. R A M. I
PROFESSORIS
REGII PROOEMIUM
mathematicum.
AD
CATHARINAM MEDICEAM,
Reginam, matrem Regis.

PARISIIS,
Apud Andream Wechelum.
1567.
Cum privilegio Regis.

Petrus Ramus (1515–72) and his *Prooemium mathematicum*.

Savile would have watched the developments across the Channel with dismay. He may also have taken an interest in the protracted debates over Ramus's will, which posthumously established a new chair of mathematics in the Collège Royal. In his stipulations for the new institution, Ramus required, in essence, that the holders of his chair should be made in his own image. They should be proficient in Latin and Greek – the minimum to be expected of any educated person, but not always achieved by some earlier professors of mathematics at Paris. Also, as Ramus had attempted to do, they should teach mathematics, not according to the usual unexamined methods but following 'reason and truth' – the ideals that Ramus himself had kept in mind as he had recast the whole of mathematics into a new form.

Savile himself, travelling in France and the rest of Europe in the late 1570s, may also have witnessed the wrangling over the appointment of the first Ramus professor, Maurice Bressieu, who underwent an examination with the stipulations from Ramus's will displayed on a brass plaque behind him.[17] Many years later, Ramus's foresight over his mathematical legacy would provide a powerful example for Savile, as he made provisions for his own vision of mathematics to be perpetuated at Oxford. Although Savile was an anti-Ramist and in his early years an enthusiastic Platonist, the influence of Ramus was a constant background throughout his life, from his first steps in mathematics and its history to the establishment of his two chairs.

The Savilian statutes

When he endowed the Savilian chairs in 1619, three years before his death, Savile drew up a set of statutes that governed the appointment of the lecturers and their duties with respect to both teaching and research. This last provision was unusual in foundations of lectureships in England, which remained close to the teaching requirements of the early-modern university. And even in the case of Savile's statutes, the research requirements remained closely integrated with his own conception of the teaching of mathematics.

Savile's foundation of his chairs finally put Oxford on a similar footing to Paris, assigning the teaching of mathematics to skilled individuals who would devote years to the task as well as to the publication of mathematical texts. He imagined that these would be young men, who could see out their entire career at Oxford as professors of mathematics: the chairs were open to any 'persons of character and repute, from any part of Christendom, well skilled in mathematics and 26 years of age'. No longer would these subjects be entrusted to the indifferent attentions of appointed regent masters and confined to ordinary lectures. Indeed, in his statutes Savile assumed that his professorships would bring the ordinary lectures to an end, with the four small stipends that the University used to pay these lecturers being divided between the two new Savilian professors.[18]

Savile stipulated that his professor of geometry was to focus his studies on the classics of Greek mathematics: Euclid's *Elements*, Apollonius's *Conics*, and the works of Archimedes. His notes on these texts would then be deposited in the new University library and kept with Savile's own books and notes, which included extensively annotated printed copies of these texts,[19] and with his astronomy colleague he could split the teaching and study of spherical geometry and trigonometry. On top of this already heavy set of responsibilities, the geometry professor was required to take on the ordinary lectures in arithmetic and music, offering instruction in theoretical and applied arithmetic, 'canonics' (the theory of music), applied geometry, and mechanics.

In a similar way, the professor of astronomy was obliged to teach the classic of ancient astronomy, Ptolemy's *Almagest*, while including in their appropriate places the discoveries of Geber, Copernicus, 'and other recent astronomers', and was also required to deposit his notes in the University library. Besides the *Almagest*, he might also teach the *Sphere*, attributed to Proclus, and Ptolemy's *Hypotheses of the Planets*, as well as astronomical calculation, all to ease the beginner into the science. In common with his geometry colleague, he had a range of additional teaching duties, and was required to introduce the students to the whole of optics, the construction of sundials, geography, and the mathematical parts of navigation. Additionally, he was forbidden to engage with astrology and the casting of horoscopes. Finally, he should make his own observations 'in imitation of Ptolemy and Copernicus', leaving these also in the library for posterity.

In these statutes, Savile had moved beyond the entirely theoretical Platonism of his youthful lectures: his professors would at least be able to demonstrate the practical applications of their subjects, and were required to teach the 'mixed' mathematical arts. Yet, as important as founding the professorships was for the future of the sciences at Oxford, one cannot help noticing how oddly Savile framed the study of astronomy in particular. Its statutes were adopted in 1619, at the end of a decade that started with the invention of the telescope and its celebrated astronomical application by Galileo, whose *Sidereus nuncius* of 1610 was as avidly read in England as anywhere else in Europe. Some of Kepler's most important works had been published earlier, including his *Astronomia pars optica* of 1604, his *Astronomia nova* of 1609, and his *Dioptrice* of 1611, while his *Epitome astronomiae Copernicanae* of 1618 and his *Harmonice mundi* of 1619 appeared in the years leading up to Savile's statutes. Galileo's revelations about sunspots had appeared in 1612 in the context of a highly entertaining and widely watched public controversy, and Copernicus's book had been condemned by the Roman Church in 1616. Finally, in 1618, the year before the adoption of the statutes, the appearance of three comets had set the astronomical world alight. The third of these, the 'Great Comet', which appeared late in the year, had provoked a flurry of telescopic observations and publications, and (in the work of well-known authors like Willebrord Snell) had dealt a near-lethal blow to Aristotelianism, the theory of solid planetary orbs, and to Ptolemaic astronomy in general.[20] In Corpus Christi College, next door to Merton, the antiquarian Brian Twyne, a generation younger than Savile, followed these events avidly, noting that 'optics' had overturned all common received astronomical and philosophical theories about the world – meaning the telescopic discoveries of Galileo and Kepler, both of whom he praised by name.[21]

In this light, Savile's insistence that his astronomy professor should focus on the *Almagest* or other ancient texts, while introducing 'modern' texts such as Copernicus's only incidentally, seems utterly anachronistic. It appears to be an attempt to perpetuate his own education and teaching in the 1560s and 1570s into a new age – even with his concessions to practical mathematics and his nod towards a style of observation (that of 'Ptolemy and Copernicus') that was already obsolete in the telescopic age.

Indeed, there is no doubt that Savile exhibited a conservative humanism, even as science had already changed far beyond what he had studied as a young master's student. While allowing that his professors might be from any nation and profession (provided that they were 'honourable'), they should nevertheless have drunk deeply from the 'springs of Aristotle and Plato' and should know at least some Greek. Even so, one should recall that Savile's new institution was intended to replace the system of ordinary lectures that went back to the foundation of the University. Perhaps he saw that the best way to ensure its success was to take over from the older system the form of commentary on authoritative texts, while always choosing (as he had in 1570) the best and most challenging texts in the genre. His desire to reform the teaching of mathematics seems to have been tempered both by his humanistic ideals and by a sense of realism born from the experience of a lifelong university administrator.

Savile's antiquarian statutes undoubtedly shaped the research interests of the early incumbents of the chairs, but the University statutes had always been flexible instruments, rather than rigid rules. For example, Savile insisted that his professors could not hold any other office – no ecclesiastical benefice, nor administration within the University or any college (despite his own wardenship of Merton), nor even a college fellowship. Yet he himself immediately appointed both his inaugural professors as fellows of Merton, and the chairs would continue to maintain a college connection, especially with New College. More importantly, even within the antiquarian framing of the positions, there was ample room for engagement with contemporary astronomical issues, as we shall see.

Oxford was not a great centre for mathematical printing. In 1520 a small pamphlet on 'compotus', the calendrical art of finding the date of Easter, was published 'ad usum Oxoniensium'. It would be over a hundred years until the next mathematical work was published – Henry Savile's *Praelectiones*,[22] a set of lectures on the first book of Euclid's *Elements*, which he had delivered in Oxford in 1620 and which were published there in the following year. Here, Savile

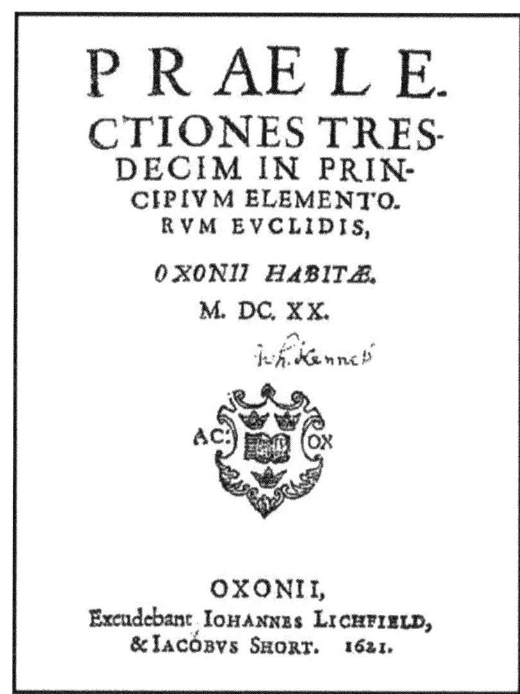
PRAELE-
CTIONES TRES-
DECIM IN PRIN-
CIPIVM ELEMENTO-
RVM EVCLIDIS,

OXONII HABITÆ.

M. DC. XX.

AC: OX

OXONII,
Excudebant IOHANNES LICHFIELD,
& IACOBVS SHORT. 1621.

Savile's introductory lectures on Euclid's *Elements* were published in 1621.

took his leave of the University, using the opportunity to survey his long career, including his meetings with famous European mathematicians and scholars and the occasional arguments that he had with them (in which he was always victorious). Then, as he said on its last page, he 'passed on the lamp' of explicating Euclid to his geometry professor.

John Bainbridge

Born in Leicestershire in 1582, John Bainbridge began his academic career at Emmanuel College, Cambridge. From 1600 to 1614 he studied there, receiving his BA and MA degrees and finally a doctoral degree in medicine. After a short while practising medicine in Leicestershire, he moved to London, where he was licensed by the College of Physicians in 1618.

Unlike Savile, Bainbridge left no student notebooks to show us how this medical doctor came to be so engaged with mathematical and astronomical questions. His early biographer, Thomas Smith, related that he had studied these sciences while bored in Leicestershire: unwilling to let an hour go to waste, he devoted himself to 'the secrets of mathematics and especially astronomy, which he had ardently loved and desired since his adolescence' – but this, like much of his biography, is entirely speculative.[23]

Smith was trying to account for the fact that this country medical doctor had, rather unexpectedly, published *Astronomicall Depiction of the Late Comet*, a learned treatise on the Great Comet of 1618, whose title page declared him to be a 'Doctor of Physicke, and a lover of the

Mathematicks'.[24] It was, moreover, a work of some historical importance. With Kepler and very few others, Bainbridge was one of the first astronomers to observe a comet through a telescope,[25] but, as one recent study has argued, his public interest in the comet seems to have been directed to the end of acquiring patronage, as requested directly in this text in his address to King James I. The treatise proved to be instrumental in his obtaining the Oxford chair of astronomy, but having secured that position, he never referred again to the promises he had made in the *Astronomicall Depiction* to publish a treatise in Latin on the comet, together with all his observations.[26]

In his treatise, Bainbridge (in agreement with Kepler and others who would publish on the same comet) showed that, based on his own parallax measurements, the comet was far beyond the sphere of the Moon. It thereby joined a long list of new celestial phenomena, starting with the new star of 1572 (Tycho's supernova), whose existence contradicted Aristotle's dictum that the heavens were eternal and unchanging. From its path along a great circle of the sky, he inferred that the comet was moving nearly in a straight line and, because it grew dimmer as it moved, it was thus clearly moving away from the Earth. Its path confirmed another of Bainbridge's anti-Aristotelian conclusions – namely, that there were no solid spheres in the heavens moving the planets. As for what the comet actually was, his estimate of its distance suggested that it was a huge body (and not a collection of vapours, as Aristotle had thought), with a diameter of about one-third of the Earth's. Following a long-popular theory, Bainbridge assumed it to be translucent, with its antisolar 'tail' being a cone of light from the Sun, refracted through its spherical mass.[27] Finally, rather coyly, he averred several times to his embrace of the Copernican system.[28]

Beyond these astronomical descriptions of the comet, Bainbridge offered an account of its *meaning*. This new celestial object, as vast as it was, could only be a direct creation of God, having been placed in the heavens as a portent, but one that (unusually in the lore of comets) did not express God's anger or herald the death of a prince. Rather, God had set it in the skies as an indication of his pleasure at the reformation of the Church, and in particular at the Synod of Dordrecht. Bainbridge read its passage through the skies as an elaborate allegory for the good fortunes of Britain and the divine favour towards James I. This was hardly 'astrology', as Bainbridge used no methods of traditional predictive astrology, but was rather (as one commentator has put it) a 'conscious and intricate game', a witty play for patronage.[29] In the event, in its ambiguity his work attracted the favour of both the astrologer Christopher Heydon and the strongly anti-astrological Henry Savile.

Bainbridge had established a connection with Savile before his comet account was even published. In early December of 1618 (when the comet was still visible in the sky), Savile wrote to him, asking for more details on his observations, and clearly this letter was written in the context of an existing acquaintance. Savile admitted that he had become less proficient in puzzling out astronomical calculations, and asked Bainbridge to provide some diagrams with his next letter, as well as details of his astronomical instruments.

Savile looked forward to conferring in person with Bainbridge directly, informing Bainbridge on his 'great busynes' that involved wrangles over land, money, and with the Court of Wards, all of which may well have been related to his establishment of the two chairs. Finally, he asked Bainbridge to pass on his regards to Henry Briggs. It is very likely that it was Briggs, then the first professor of geometry at Gresham College in London, who had first brought Bainbridge to Savile's attention.

Bainbridge's first act as Savilian professor was to publish in 1620 an edition of the Greek text (with Latin translation) of two ancient works on elementary astronomy. In his dedicatory

John Bainbridge (1582–1643), the first Savilian professor of astronomy.

epistle to the third Earl of Pembroke, chancellor of Oxford University, Bainbridge celebrated the flourishing of all the arts under the Earl's direction, including mathematics which Savile had 'called forth from the dust and the tomb'.[30] Bainbridge and his fellow mathematical professor now asked for Pembroke's patronage too.

In his dedicatory epistle, Bainbridge also focused on Savile's newly established professorships. Accurately echoing their founder's intentions as expressed in his statutes, Bainbridge wrote that Savile believed that his money would be well spent on the chairs if the ancient mathematicians themselves could be enthroned upon them, in the form of new editions based upon the ancient manuscripts. While his companion Henry Briggs had set himself the task of editing Euclid's *Elements* and had published an edition of Books I–VI in that same year, Bainbridge had undergone the perhaps unequal task of Ptolemy's *Almagest*. In order to prepare a way to his edition, he had chosen two smaller tracts: *De sphaera* (On the Sphere), which had long been attributed to Proclus, but which Bainbridge recognized as really an extract from Geminus, and the extant Greek text of Ptolemy's *De planetarum hypothesibus* (On the Planetary Hypotheses). As we saw above, both texts were mentioned in Savile's statutes as texts which could be taught alongside the *Almagest*, and it is very likely that Savile included these works already knowing of Bainbridge's work on them.

Moreover, in publishing the *Sphere*, Bainbridge was continuing a long interest both of Savile himself and of Savile's younger brother Thomas, who had died nearly three decades earlier at a relatively young age. In the early 1580s Henry Savile had worked on the text of Geminus's *Elements of Astronomy* (from which the *Sphere* was excerpted) as he travelled through Europe. First, he and his travelling companion George Carew copied out the work in Vienna, and then, while he was resident at Gian Vincenzo Pinelli's library in Padua, Pinelli had two fair copies made of the Greek text; another copy was also made by András Dudics. When Thomas Savile followed much the same route through Europe in the late 1580s, he made a Latin translation of the text while staying with Dudics; Pinelli also made copies of this translation once Thomas reached Padua. Furthermore, while Thomas was still on the Continent, engrossed in the study of Geminus, his brother provided the manuscripts for Edo Hilderich for his *editio princeps* of the Greek text and the first published Latin translation (in which Henry Savile is gratefully acknowledged); Bainbridge's text was actually the last printed translation of the excerpt from Geminus that went under Proclus's name.[31] In publishing the *Sphere*, Bainbridge had made available to Oxford students an elementary astronomical text with an ancient pedigree (unlike that of the ubiquitous Sacrobosco); he also put his name to a text that for many years had preoccupied both the founder of the chair and his late brother.[32]

The other text, Ptolemy's *Planetary Hypotheses*, while less connected to Savile's own scholarly interests, was of greater historical significance. In the first book of this work, Ptolemy had set out an account of the cosmos that carefully nested the spheres of the planets within each other, containing in each the complicated mechanisms of deferents and epicycles, and turning the mixture of models in the *Almagest* into the 'Ptolemaic system'. This first book ends with a series of calculations, setting out the distances and sizes of all the planets, while the second book is more philosophical in tone, first attacking the theory of nested spheres found in Aristotle's *Metaphysics*, and then showing that the actual mechanisms of the planets might be physically contained within thick bands, rather than spheres encompassing the entire cosmos.[33]

Much of this was already known indirectly through medieval Arabic astronomers, and informed the medieval and early-modern *Theorica planetarum*. Only the first part of Book I, without the final calculations of distances and sizes, was still extant in Greek, and this was the work that Bainbridge published in 1620, establishing it (as he records in his prefatory remarks) from three very corrupt manuscripts. This first printed edition of the Greek text was completed to a very high quality, as was the Latin translation (again, the first to be made). Both Heiberg's authoritative edition and the most recent study and translation of the Greek text draw on Bainbridge's emendations, particularly of the numerical parameters of the spheres and models, while the historian of astronomy Noel Swerdlow, a critical reader of past scholarship, described both the edition and its translation as 'excellent'.[34] Bainbridge was indeed ably fulfilling the Savilian role of the erudite humanist mathematician.

Yet, he narrowly missed out on translating the *whole* of Ptolemy's work on the heavenly spheres. After his interests turned to Arabic texts, Bainbridge wrote to the great Leiden orientalist and mathematician Jacobus Golius, following up on some Arabic texts that Golius had mentioned in a published list of manuscripts that he acquired in his travels. But somehow, Bainbridge missed Golius's reference to an Arabic version of the *Planetary Hypotheses*, while Golius owned one of only two extant manuscripts of the Arabic translation but did not understand its significance. Bainbridge too overlooked it, and it would be brought to light again only three centuries later, in Heiberg's edition of the entire text.[35]

For much of his career, Bainbridge turned his attention to continuing the Savilian project with a new edition and translation of Ptolemy's *Almagest*. His ambitions went beyond simply correcting the Greek text. The *Almagest* was no mere antiquarian text, but was to form the basis of astronomical teaching at Oxford, even though modern astronomers, such as Tycho Brahe and Kepler, had repeatedly shown that its planetary predictions were out of step with the best observations. Thus, Bainbridge embarked upon what may seem to us to be a quixotic project of improving the text, by remeasuring the latitudes of Rhodes and Alexandria (from

A portrait of John Bainbridge and his daughter, painted by Peter Lely around 1640.

where Hipparchus and Ptolemy had made their observations) and establishing longitudes by the simultaneous observation of lunar eclipses. He was also convinced that the lost works of these ancient astronomers, essential for restoring the *Almagest*, were either languishing in European libraries or could be recovered from Arabic sources. He asked travellers on the Continent to search out these texts, while John Greaves, his protégé and successor to the Savilian chair of astronomy, sought out manuscripts for him in the East.[36]

Bainbridge's *Almagest* project led him to study Arabic himself. In 1626 he knew only how to read the mathematical tables in Arabic astronomical manuscripts, but not the text itself, as he informed Archbishop Ussher in a letter in which he vowed to learn the language well enough that he could 'adorn' and 'enrich' his *Almagest*. As he embarked on these studies, he was already benefiting from the friendship and guidance of England's greatest Arabist, William Bedwell.[37] It was perhaps through Bedwell's connections at Leiden that Bainbridge himself tried unsuccessfully to forge scholarly relationships with orientalists there, including Thomas Erpenius and Golius.

The *Almagest* project was never completed. It may be that the fantastically complicated task of determining the correct dates of astronomical events waylaid him, for his final project was more directly concerned with issues of chronology.[38] His expertise in this area was put into the service of Archbishop Ussher's own work on biblical chronology, and he corresponded with Ussher frequently.[39]

In his own career, Henry Savile had involved himself in controversy with one of the greatest European scholars, Joseph Scaliger. Savile, along with many other European mathematicians, took up arms against Scaliger's disastrously conceived vanity project, the *Cyclometrica* of 1594,

Cl. V. Iohannis Bainbrigii,
Aſtronomiæ,
In celeberrimâ Academiâ Oxonienſi,
Profeſſoris Saviliani,
Canicularia.

Unà cum demonſtratione
Ortus *Sirii* heliaci,
Pro parallelo inferioris Ægypti.

Auctore Iohanne Gravio.

Quibus acceſſerunt,
Inſigniorum aliquot Stellarum Longitudines, & Latitudines,
Ex Aſtronomicis Obſervationibus
Vlug Beigi,
Tamerlani Magni nepotis.

Oxoniæ,
Excudebat Henricus Hall, Impenſis
Thomæ Robinson, 1648.

(*Left*) John Bainbridge acquired several Arabic and Persian books, including this manuscript of the 15th-century Timurid Sultan astronomer Ulugh Beg. (*Right*) Bainbridge's *Canicularia* was one of the earliest Oxford books to use Arabic type.

which was an absurd but superficially impressive quadrature of the circle.[40] Bainbridge tackled Scaliger on his most secure territory, chronology and its emendation. One of Scaliger's most impressive claims had been the invention of time reckoning by the Julian date, extrapolating Julius Caesar's calendar back to a starting point of 1 January in the year 4713 BC.[41] Each day in history since could then be assigned a day number, and eclipses and other astronomical events could be used to match these uniquely identified days to the dates provided in the bewildering range of calendars used throughout the ancient world. Astronomers use this chronological method to this day.

Bainbridge found many errors in Scaliger's calibration of the Egyptian calendar against the new uniform calendar, for which he used ancient observations of the rising of Sirius, the most important astronomical event in ancient Egypt. Bainbridge had not completed this work by the time of his death in 1643, but John Greaves, his immediate successor to the Savilian chair, published it posthumously in 1648 as the *Canicularia*. Although still little studied, it has been described by the leading modern authority on Scaliger's chronology, as 'the most original of all 17th-century works on ancient chronology', which 'demolished central portions' of Scaliger's work on the Egyptian calendar.[42] Greaves composed the dedicatory poem that adorned Bainbridge's lavish monument in Merton College Chapel, which records that he emended Scaliger more successfully than Scaliger himself could emend chronology.[43]

2		Logarithmi.
1	0	0000,00000,00000
2	0	3010,29995,66398
3	0	4771,21254,71966
4	0	6020,59991,32796
5	0	6989,70004,33602
6	0	7781,51250,38364
7	0	8450,98040,01426
8	0	9030,89986,99194
9	0	9542,42509,43932
10	1	0000,00000,00000

Some of Henry Briggs's logarithms to base 10, calculated by hand to fourteen decimal places.

One recent scholar's verdict on Bainbridge – that he 'achieved very little by his pursuit of Arabic'[44] – is perhaps not entirely unjustified. In his work on the *Planetary Hypotheses*, he narrowly missed the opportunity to publish the entire text from the Arabic translation. His years of work on Arabic astronomical sources did not eventuate in a new version of the *Almagest* as he had originally intended, and his chronological work, unfinished at his death, saw the light only posthumously in a brilliant (but almost entirely neglected) publication. However, his work prepared the way for his successor John Greaves and also for later Savilian professors of both geometry and astronomy. Within the rather narrow confines of Savile's statutes, which prioritized the study of Latin and Greek texts from antiquity, Bainbridge both demonstrated the value of Arabic learning and showed ways in which Savile's model of the philological study of mathematics and astronomy could be brought up to date, contributing to contemporary debates and stimulating a programme of observations.

Henry Briggs and Peter Turner

Bainbridge's companion, the Savilian Professor of Geometry Henry Briggs, also balanced antiquarian pursuits with the practice of mathematics, and in the latter he was more notable than Bainbridge. Much of his mathematical work during his tenure was devoted to the calculation of logarithms. Briggs was responsible for devising the notion of logarithms to base 10, and published extensive tables of these for the use of calculation – work that he had started in

London while Gresham professor of geometry in the years before his election to the Savilian chair. These tables, completed by the Dutch mathematician Adriaan Vlacq, remained in use well into the 18th century. His work on trigonometry was published posthumously by his student Henry Gellibrand as the *Trigonometria Britannica*; it included a table of the logarithms of sines and tangents, greatly simplifying the work of astronomers and others. As a practitioner, Briggs seemed 'modern' in his sensibilities, preferring the Copernican system to the Ptolemaic one, and praising Galileo's discovery of the moons of Jupiter.[45]

As a scholar of ancient texts, Briggs continued the task handed to him by Savile in his inaugural lecture, producing an edition of the first six books of Euclid's *Elements*; these books continued to form the core of elementary instruction in geometry well into the 20th century. Although Briggs planned a complete edition of all thirteen books, his choice to publish only the first six was clearly motivated by pedagogical needs and (apparently) by his difficulty in finding a publisher for the complete text.[46]

After his death in 1631, Briggs was succeeded as Savilian Professor of Geometry by Peter Turner, who, like Briggs, had also held the Gresham chair in the same subject. While an excellent linguist, he seems to have contributed very little to mathematics during his tenure of the Savilian chair. Devoting himself more to university administration than to scholarship, his career foundered during the chaos of the Civil War. In 1649, three years before his death, he was succeeded in the chair by a mathematician of an altogether different calibre, John Wallis.[47]

John Greaves

As mentioned earlier, Bainbridge's successor as Savilian Professor of Astronomy was his student John Greaves, who had also been a Gresham professor of geometry.[48] His tenure of the chair was, like Turner's, cut short by the reforms of the Parliamentary visitors, who dismissed him in 1648 after just five years in the post. Nevertheless, as one scholar has written, Greaves managed to achieve 'what John Bainbridge, his mentor, could not achieve for himself'.[49]

Benefiting both from Bainbridge's instruction and from the many travels he made through the Levant in search of manuscripts for his teacher's *Almagest* project, Greaves attained a profound knowledge of languages, manuscripts, and mathematics. He travelled through Egypt in 1637, searching out manuscripts and correcting Ptolemy's latitude measurements for his places of observation.

During that trip, he became fascinated by the pyramids and explored the passages within the Great Pyramid of Cheops. He wrote an account of this adventure, and a description of the inner and outer dimensions of the pyramid in his *Pyramidographia* of 1646, a masterpiece of description and his best-known work. He continued his teacher's concern with chronology, sifting through conflicting historical accounts and astronomical records in order to date the building of the pyramid, although he calculated a much more recent date than is now generally accepted.[50] Apart from this work, Greaves dedicated himself to publishing important astronomical tables from the 15th-century observatory of the Timurid sultan Ulugh Beg, and also published a work on chronology that made use of these observations. Written in Persian, these works also led him to produce the first Persian grammar to be published in England.

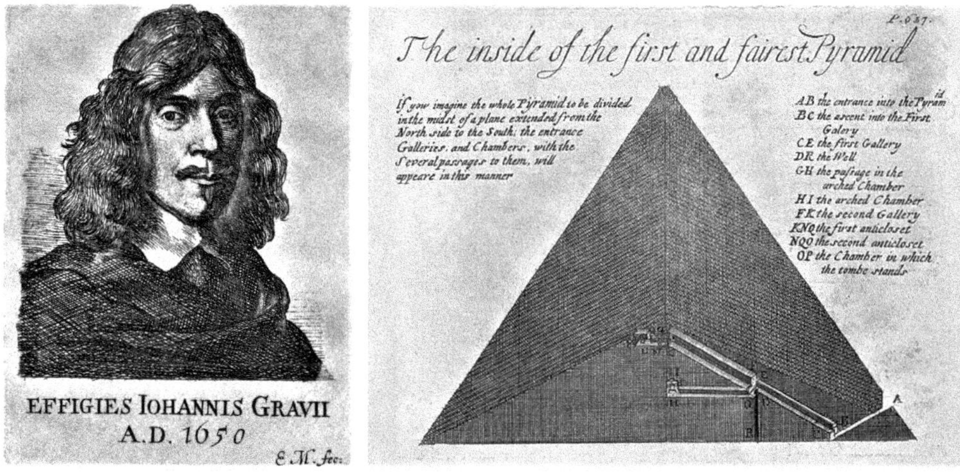

John Greaves (1602–52), the second Savilian professor of astronomy, and a diagram from his *Pyramidographia*.

Conclusion

Savile's founding of the chairs of astronomy and geometry set the study of the sciences at Oxford down a particular course, with two important features. First, because these professorships were designed to replace the old regent system of teaching astronomy and geometry, there was a focus on the elementary texts of the university curriculum – Euclid and Ptolemy, in particular. In the case of the astronomy chair, this set the teaching of the science at Oxford rather at odds with the rapid developments in astronomy that were taking place, especially on the Continent.

Hand in hand with this went an emphasis on the editing and publishing of ancient texts. 'Scientific antiquarian', a description that one modern scholar has applied to Greaves, could indeed characterize all the Savilian professors in this period.[51] Yet even within the rather narrow focus of these professors, we find some innovations that Savile had not anticipated. First and foremost was their growing interest in Arabic and Persian as a means of rectifying and recovering ancient texts and accessing the subsequent astronomical work, especially of Eastern scientists. They also begin to evince an engagement – perhaps still at arm's length – with more recent developments in the sciences, and to contribute their observations to the works of other better-known mathematicians.

The Civil War marked a break in both the chairs. Yet their successors, many of them far greater mathematicians and scientists, would continue to build upon the early professors' legacy, making use of their books and manuscripts and in many cases completing projects that they had left unfinished.

Seth Ward (1617–89), the third Savilian professor of astronomy.

CHAPTER 2

Interregnum and restoration

ADAM JARED APT

In view of the physical and political dislocations and disturbances of Oxford leading up to the interregnum followed by the stresses of that period, the resilience of the Savilian chair in astronomy during those years can be credited to its holder, Seth Ward, who occupied the position from 1649 to 1660. The two succeeding holders, Christopher Wren (from 1661 to 1673) and Edward Bernard (from 1673 to 1691), were men of high calibre, although only the former is widely recognized, and not for his work as Savilian professor. During the tenure of both, there was some laxity in the continuance of the work mandated by the Savilian statutes.

Seth Ward

Seth Ward, born in 1617, entered Sidney Sussex College, Cambridge, in 1632, where he proceeded to BA and MA degrees in 1637 and 1640.[1] According to his friend and biographer, Walter Pope, Ward attained his proficiency in mathematics from books in the college library, and 'when he was Sofister, he disputed in those Sciences, more like a Master than a Learner, which Disputation Dr. Bainbridge heard, greatly esteemed, and commended'.[2]

Two of Ward's contemporaries at Cambridge were John Wallis, who became Ward's colleague as Savilian professor of geometry, and Jeremiah Horrocks, whose work came to prominence only in the 1660s, after his death, and yet there is no evidence that Ward, Wallis, and Horrocks knew each other as students.[3] Ward seems to have learned through self-study rather than through engagement with the mathematicians at Cambridge. In particular, he made a close study of the first edition of William Oughtred's classic text, *Clavis mathematicae* (The Key to Mathematics) of 1631, going so far as to visit him, along with his friend Charles Scarburgh, for further instruction. According to Oughtred:[4]

> But occasion was administered by one Mr. Seth Ward, a young man excellently accomplished with all parts of polite Literature . . . who took pains to seek me out at my house, and by a gentle violence induced me to publish again my former Tractate in a manner new moulded and perfected.

Adam Jared Apt, *Interregnum and restoration.* In: *Oxford's Savilian Professors of Astronomy.* Edited by: Robin Wilson and Steven Balbus, Oxford University Press. © Oxford University Press (2025). DOI: 10.1093/oso/9780198894292.003.0002

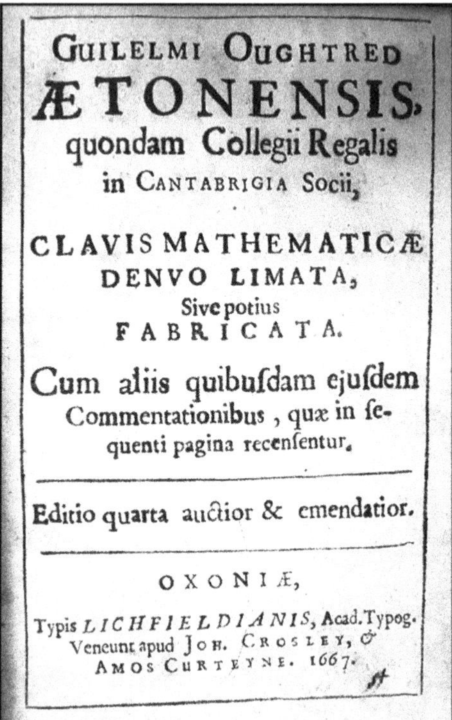
GUILELMI OUGHTRED
ÆTONENSIS,
quondam Collegii Regalis
in CANTABRIGIA Socii,

CLAVIS MATHEMATICÆ
DENVO LIMATA,
Sive potius
FABRICATA.

Cum aliis quibuſdam ejuſdem
Commentationibus, quæ in ſe-
quenti pagina recenſentur.

Editio quarta auctior & emendatior.

OXONIÆ,

Typis LICHFIELDIANIS, Acad. Typog.
Veneunt apud JOH. CROSLEY, &
AMOS CURTEYNE. 1667.

William Oughtred (1575–1660) and an edition of his *Clavis mathematicae*.

Ward and Scarburgh then used the book as a textbook for Cambridge students. Ward was elected the University's mathematical lecturer in 1643.[5]

Apart from some non-specific attestations to his mathematical sagacity, the above summary of Ward's education constitutes nearly all that we know of his qualification for appointment to the Savilian chair. Although it appears scant in comparison with the early work of other occupants, Ward may have done astronomical work of which all record is lost, or there was a determination to recruit a mathematician for the chair in astronomy.

To appreciate Ward's appointment to the Savilian chair, we must note his religious and political views, because his changing allegiances provoked some contemporaries to judge him harshly. Aligned against the Puritans, Ward was among the Cambridge scholars who resisted parliamentary intervention in the University. His first publication was a collaborative work in 1644 with six other authors (including Isaac Barrow, who would become Cambridge's first Lucasian Professor of Mathematics), opposing the Solemn League and Covenant, and when Ward refused to sign it, the Parliamentary visitors promptly ejected him from the University. At this difficult time he did not lack for friends, and one after another (including William Oughtred) took him in.

Ward's appointment

In Oxford John Greaves, ejected from his Savilian chair by the Parliamentary visitors, was determined to have some say in the choice of his successor, and to that end sought the advice of other Royalists. These included Scarburgh, who had become a prominent London physician, and William Holder, another close friend of Ward's from Cambridge;[6] both men would also influence the career of the next Savilian professor of astronomy.

Greaves had heard of Ward but had never met him, and Scarburgh and Holder convinced him that Ward was the man best suited to succeed him. Soon after, knowing nothing of this and visiting Oxford, Ward had a chance encounter with Greaves, who introduced himself. As Ward's biographer Walter Pope later recalled:

> I remember I have heard the Bishop [Ward] say, that amongst other arguments, Mr. Greaves told him, if you refuse it, they [the visitors] will give it to some Cobler of their Party who never heard the name of Euclid, or the Mathematics, and yet will greedily snap it up for the Salaries sake.

Ward was naturally pleased with Greaves's offer, but pointed out that as he had been dismissed from Cambridge for refusing to subscribe to the Covenant, he would surely never be appointed. Greaves replied that he and friends had already taken this into account, and he arranged a fix with Sir John Trevor,[7] who had influence with the appointment committee. Although a parliamentarian, Trevor was a moderate and supportive of the best scholars, regardless of politics. Ward duly acceded to the Savilian chair of astronomy in 1649 and dedicated to Trevor the *De cometis* of 1653, one of his first two astronomical publications; it contains one of the earliest suggestions that cometary orbits might be elliptical.[8]

The Oxford in which Ward arrived was far from idyllic. During its occupation as the Royalist capital during the Civil wars, the city had become overcrowded, squalid, disease-ridden, and dangerous, and both the town and the University must have remained so as Ward took up his position. Furthermore, the University's politics were fraught, as Oxford had been largely on the losing side of the wars. Not all who supported the King had been ejected from their posts, but those remaining had to hold their tongues and negotiate University and college affairs with care.[9]

Controversy over Ward's consistency of principle began in early 1650, and initially concerned whether, under penalty of ejection from his college and despite his having refused to subscribe to the Solemn Covenant, he now subscribed to the Engagement, which stated:

> I do declare and promise, that I will be true and faithful to the Commonwealth of England as it is now established, without a king or House of Lords.

We may well imagine the pressure on him to conform, given that the alternative would involve forfeiting the most congenial position for someone with his interests.[10]

Ward chose to become a fellow of Wadham College, a happy (and probably a deliberate) choice, because this brought him close to John Wilkins, whom the Parliamentary visitors had appointed Warden the year before, and with whom he developed a productive and professional relationship. (Wilkins and Walter Pope were half-brothers.) Further, Christopher Wren arrived at Wadham as an undergraduate in the next spring. According to John Wallis, who had taken up the Savilian chair in geometry a few months before Ward took up the chair in Astronomy:[11]

> About the year 1648–49, some of our Company were moved to Oxford, first Doctor Wilkins, then I, and soon after, Doctor Goddard. Doctor Ward, Doctor Petty and many others of the most inquisitive persons met at Oxford at Doctor Petty's lodgings so long as Doctor Petty continued at Oxford, and for some time afterward.

John Wilkins (1614–72), Warden of Wadham College, and John Wallis (1616–1703), Savilian Professor of Geometry.

Wallis went on to say that they then moved their meetings to Wilkins's lodging at Wadham.

This particular Oxford group was a precursor to the Royal Society of London. Having started with Royalist sympathies, and then leaning towards the Parliamentary cause while remaining a moderate in religion and politics, Wilkins was even more adept than Ward at walking the tightrope between the two factions, and Wadham became both a limited refuge from the politics of the University and a meeting place for the like-minded in the sciences.

Earlier historians have portrayed 17th-century Oxford and Cambridge as oblivious or hostile to new developments in the mathematical sciences, and as stuck in the medieval past, with most creative work being done elsewhere, but this is far from the truth. Much of this arises from misleading polemical writings of contemporaries, like Thomas Hobbes, who had an axe to grind. In reality, the two universities were populated with scholars who were up to date with the latest thinking and discoveries, taught them to their students, and pursued original research.[12]

Seth Ward was part of this story. According to Pope, Ward's first task was to revive the obligatory Savilian lectures, which had been neglected during the wars, and as far as we can determine, he lectured diligently as required by the Savilian statutes; indeed, he claimed never to have missed a scheduled day. (His lecture notes have not survived.) He tutored students gratis, and even foreigners who came to Oxford, and he expected his students to be serious in their studies.

Ward also participated actively in the current astronomical developments in England and on the Continent, and under his tutelage and with the others of Wilkins's circle, Wren began astronomical observations. Ward also encouraged Robert Hooke, who had come up to Oxford as an undergraduate in 1653 or 1654, to take up the study of astronomy, to such good effect

Under the wardenship of John Wilkins, Wadham College became the centre of Oxford experimental philosophy in the 1650s. In this engraving of 1738 Wilkins is at the front (showing an image of Gresham College), with Seth Ward, Christopher Wren (displaying the Sheldonian Theatre), and others. The tower over the entrance contained Ward's astronomical observatory.

that Hooke soon produced astronomical instruments with which he made observations that he later presented to the Royal Society.[13]

Ward was now a colleague of John Wallis, which must have been a further intellectual stimulant. Although Wallis was a Presbyterian and a supporter of Parliament, it appears that such differences were irrelevant and that the two had a cordial relationship. In spite of a reported tiff between them over who had seniority when they proceeded to their Oxford degrees in Divinity there seems to have been no animosity between them.[14]

The Oxford circle around Wilkins grew to include such leading figures in the mathematical and natural sciences as William Petty, Robert Boyle, Sir Paul Neile (whose son William matriculated as a Wadham undergraduate in 1652), Robert Moray, Thomas Willis, Jonathan Goddard (physician and Warden of Merton College from 1651, to whom Ward incorrectly attributed the first English construction of a telescope),[15] and Ward's close friend from Cambridge, the mathematician and astronomer Lawrence Rooke, who left Wadham in July 1652 to take up the Gresham chair of astronomy in London.[16]

Early in Seth Ward's tenure, and in conformity with the Savilian statutes' requirement to engage in and promote observation, he set up an astronomical observatory at Wadham that was used assiduously by Wren. It seems to have been located in the tower over the entrance to the college,[17] and Ward lived just below it, in 'the chamber over the gate'.[18] As Ward wrote at the time:[19]

> I have spent much of my time . . . in building a slight observatory for the matter of my profession and in procureing & fitting Telescopes and other instruments for observation.

This effectively shifted the University's astronomical activities from the top storey of the Schools quadrangle's tower, which continued to house its other astronomical instruments.

The observatory was fitted out with a telescope constructed by Richard Reeve,[20] a highly regarded London optician.[21] Writing in 1655, Samuel Hartlib reported that Wren 'counteth Reeve makes the best microscopes of any to be had. As likewise tubes of which they have one at Oxford of 24 foot long, which with a thread once placed hee can manage as he pleases'. This telescope was dubbed 'the Oxonian', and Wallis seems to refer to it in a letter to Huygens about his observations of Saturn, as well as to telescopes of 12, 22, and 28 feet in length.[22] We have no records of Ward's own observations at the observatory, and nor does his name appear in any of the accounts of observing sessions there (unlike those by the young Christopher Wren), and it may be that he was not a systematic observer. Even William Oughtred, requesting Oxford observations of the comet of December 1652, wrote to Lawrence Rooke rather than to Ward, but it was Ward who replied cordially to his old friend, enclosing Rooke's observations.[23]

The Boulliau–Ward debate

Ward's most memorable contribution to the practice of astronomy, and the one that attracted the greatest attention among his contemporaries, arose from a dispute with the French astronomer Ismaël Boulliau over how best to forecast planetary positions. The background to this issue was Johannes Kepler's introduction in 1609 of two of his laws of planetary motion, that the planets' orbits around the Sun are elliptical with the Sun at one focus, and that a line connecting a planet to the Sun sweeps out equal areas in equal times.

Kepler had introduced a new celestial physics to explain planetary motion, but it was not generally accepted, except for its idea that a force based in the Sun had something to do with the motion. At the time of Ward and Boulliau astronomy remained primarily kinematic (as it had been since antiquity), at least in practice. This produced a problem, sometimes called 'solving Kepler's equation', for which there is no exact analytical solution.[24]

The problem is deceptively simple, and asks for the angle between an orbiting planet and a starting position, given that starting position, the eccentricity of the ellipse, the mean angular speed of the planet in its orbit, and the elapsed time. Although Kepler and others had solved it, the issue remained that, from the perspective of 17th-century astronomers, their solutions were not mathematically elegant. A partial reason for this was that, as Kepler suspected but did not prove, the problem was not susceptible to an exact analytical solution – and so began the attempts to find 'purer' solutions that might additionally be easier to calculate than an approximation.[25]

The Boulliau–Ward debate began with Boulliau's publication of his *Astronomia philolaica* in 1645, when Ward was not yet at Oxford. Boulliau accepted that planetary orbits were ellipses

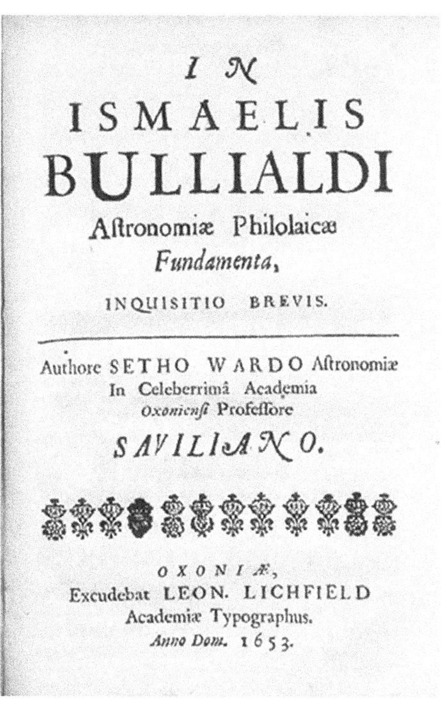

I N

ISMAELIS

BULLIALDI

Aſtronomiæ Philolaicæ

Fundamenta,

INQUISITIO BREVIS.

Authore SETHO WARDO Aſtronomiæ
In Celeberrimâ Academia
Oxonienſi Profeſſore

SAVILIANO.

OXONIÆ,
Excudebat LEON. LICHFIELD
Academiæ Typographus.
Anno Dom. 1653.

[*Left*] Ismaël Boulliau (1605–94). [*Right*] Seth Ward's attack on Boulliau's method.

with the Sun at one focus, but he rejected Kepler's physics and derived the ellipses from a complicated argument that was based in part on a reversion to the classical ideal of uniform circular motion of celestial objects. His work was problematic, not least for replacing Kepler's area law with something more complicated.

The first attack on Boulliau's method was Ward's Oxford 1653 tract *In Ismaelis Bullialdi astronomiae philolaicae fundamenta, inquisitio brevis*. Ward praised Boulliau's effort, but showed that Boulliau's calculational methods were inconsistent with each other and with their premises. In 1657 Boulliau replied, modifying his argument somewhat along the lines suggested by Ward, so the calculated positions now agreed more closely with Kepler's own calculations. But he then reasserted his pre-Keplerian physics, in contrast to Ward who, in his *Astronomia geometrica* of 1636, had confirmed his own belief that Kepler correctly hypothesized that the Sun was 'the true and physical instrument of the planetary motions'. But the combined methods of Ward and Boulliau still had the disadvantage of replacing Kepler's area law with one that less accurately predicted actual planetary positions.[26]

The full story extends beyond Boulliau and Ward, and this episode is part of the larger story of celestial physics between Kepler and the great Newtonian synthesis. But Ward's modified version of Boulliau's methods took hold in England, as in John Newton's 1657 textbook *Astronomia Britannica . . . As it is illustrated by Bullialdus, and the easie way of Calculation, lately published by Doctor WARD*; this textbook, in straightforward English, was intended for use in schooling navigators. It was probably also for use in the universities, but would have been problematic for Oxford because it introduced decimal angular measures whereas the Savilian statutes required instruction in sexagesimal calculation.[27]

Ward's method was also used by Thomas Streete in 1661 in his *Astronomia Carolina*, the textbook from which Isaac Newton learned of Kepler's laws of planetary motion.[28] But it was a dead end, being significant in retrospect mainly for facilitating the acceptance of Kepler's astronomy by making it more easily practicable.[29]

Ward and theology

Until the late 18th century, familiarity with the history of astronomy was requisite for the active study of astronomy, and Seth Ward showed signs of an antiquarian interest in the subject. He was Aubrey's source for a story that the mathematician Thomas Harriot denied the biblical account of the Creation, and for an anecdote that found poetic justice in the cancer that killed him as retribution for his supposed atheism.[30] Although Aubrey related this with his characteristic light touch, Ward, because of his religious convictions, may have relayed the story to Aubrey with somewhat more gravity. Ward was also Aubrey's source for the familiar anecdote about Henry Savile's rejection of Edmund Gunter as a candidate for the position of first occupant of the chair in geometry.[31]

Most importantly, Ward encouraged or commissioned Thomas Hyde's Latin translation of the star tables of the 15th-century Samarkand astronomer Ulugh Beg, which Hyde, as Bodley's Librarian, eventually published in 1665. This star catalogue thereby became a commonality of the first three Savilian professors of astronomy, having been collected by Bainbridge (see Chapter 1), edited in part by Greaves (who published excerpts with Bainbridge's posthumous *Canicularia* in 1648), and then further translated at the instigation of Ward.[32]

While occupying the astronomy chair, Ward also devoted time to theology; this makes sense of his later move to an ecclesiastical career after his Savilian position. Throughout his time in Oxford, he delivered sermons at St Mary's, the University church, although not required to do so by the Savilian statutes. Indeed, it was mainly as a theologian that Ward became a participant in the long-running feud between John Wallis and Thomas Hobbes over Hobbes's claims to have solved the ancient geometrical problem of squaring the circle. Although the story of this feud is more about Wallis than Ward, it greatly occupied Ward's attention during his Oxford years and helps to illuminate his philosophy and his views on university education. This dispute went beyond the simple matter of a gross mathematical error; it was also about the nature of mathematics, epistemology, and theology.[33]

Ward and Hobbes were initially friendly, and Ward may have contributed the favourable 'publisher's preface' to Hobbes's *Humane Nature* in 1650. But a turn for the worse came in 1651 with the publication of Hobbes's *Leviathan*, in which his pure materialism, hardly separable from the rest of his religious heterodoxy, now lay exposed to view. Hobbes's materialism did not allow for the immortality of the soul: this offended Ward, who almost immediately published a rebuke. So although Ward was a member of the Oxford circle that was associated with promoting the 'mechanical philosophy', this philosophy did not necessarily coincide with thoroughgoing materialism.

The next stage in Ward's conflict with Hobbes was precipitated by John Webster, a radical figure who in 1653–54 levelled a broadside at the two universities in his *Academiarum examen, or the Examination of Academies*.[34] Ward and Wilkins viewed this as a serious threat, possibly because it might influence the more radical Parliamentarian reformers in London, and immediately responded anonymously with a polemical counterblast, *Vindiciae academiarum*,

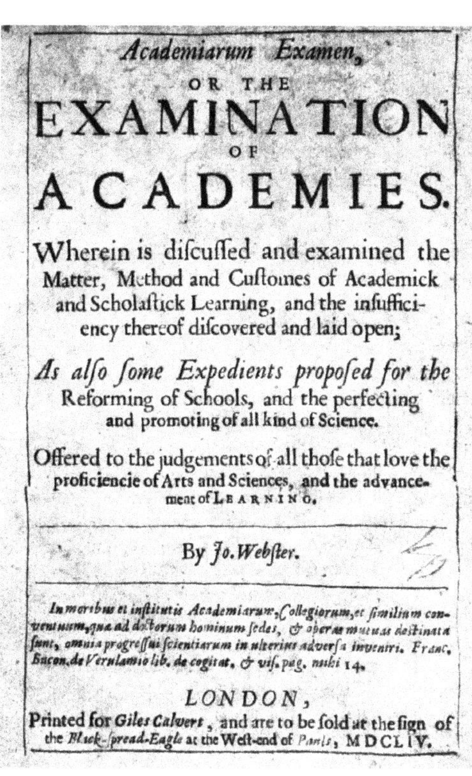

Academiarum Examen,

OR THE

EXAMINATION

OF

ACADEMIES.

Wherein is difcuffed and examined the Matter, Method and Cuftomes of Academick and Scholaftick Learning, and the infufficiency thereof difcovered and laid open;

As alfo fome Expedients propofed for the Reforming of Schools, and the perfecting and promoting of all kind of Science.

Offered to the judgements of all thofe that love the proficiencie of Arts and Sciences, and the advancement of LEARNING.

By *Jo. Webfter.*

In moribus et inftitutis Academiarum, Collegiorum, et fimilium conventuum, quæ ad doctorum hominum fedes, & operas mutuas deftinata funt, omnia progreffui fcientiarum in ulterius adverfa inveniri. Franc. Bacon. de Verulamio lib. de cogitat. & vif. pag. mihi 14.

LONDON,

Printed for *Giles Calvert*, and are to be fold at the fign of the *Black-fpread-Eagle* at the Weft-end of *Pauls*, MDCLIV.

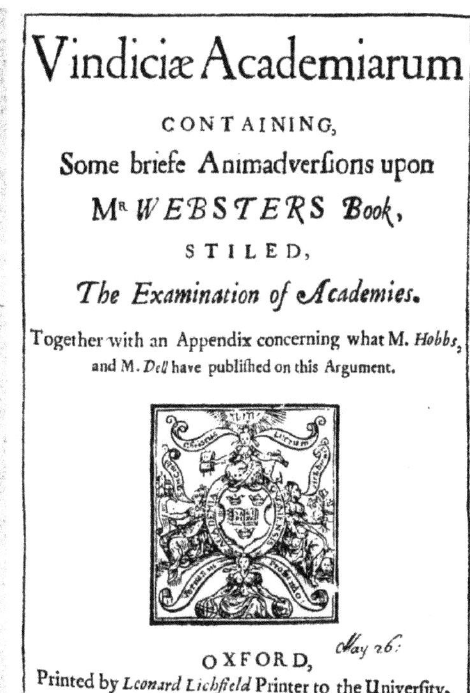

Vindiciæ Academiarum

CONTAINING,

Some briefe Animadverfions upon

MR *WEBSTERS Book*,

STILED,

The Examination of Academies.

Together with an Appendix concerning what M. *Hobbs*, and M. *Dell* have publifhed on this Argument.

OXFORD,

Printed by *Leonard Lichfield* Printer to the Univerfity, for *Thomas Robinfon*. 1654.

[*Left*] John Webster's attack on the universities for their neglect of practical subjects and ignorance of recent developments in mathematics and astronomy. [*Right*] Seth Ward vigorously rebutted Webster's 'buzzing Discourse', the 'puerility of his Arguments', and his 'crude and jejeune Animadversions'.

Containing Some Briefe Animadversions upon Mr Webster's Book, with Ward writing the main text and Wilkins the preface. It made broad use of ridicule, sarcasm, parodic writing in the style of Webster's grandiloquent and esoteric language, accusations of plagiarism, and charges of misunderstanding quoted texts. Refuting Webster, who accused the universities of being hidebound and suggested the introduction of Copernicus and Francis Bacon as well as Rosicrucianism and Paracelsian chemistry, Ward showed that Copernicus and Bacon were both taught, with Copernicus's theory in its Keplerian form. Moreover, although Aristotle had been superseded in astronomy and natural philosophy, he still had a proper place in the university curriculum as one of the greatest philosophers. Other authors whom Webster recommended (Ward claimed) were merely 'mysticall' and had no place in a university education. In his conclusion, Ward asked rhetorically:

> Of those very great numbers of youth, which come to our Universities, how few are there whose designe is to be absolute in Naturall Philosophy?'

He answered that they may go on to the Inns of Court, and that the point of their education was that 'their reason, and fancy, and carriage, be improved . . . that they may become Rationall and Gracefull speakers, be of an acceptable behaviour in their Countries [that is, the provinces]'.

In response to Webster's charge that the universities continued to teach Ptolemaic astronomy, Ward wrote:

> I believe there is not one man here [at Oxford], who is so farre Astronomicall, as to be able to calculate an Eclipse, who hath not received the Copernican system, (as it was left by him, or as improved by Kepler, Bullialdus, our own Professor [referring to himself], and others of the Ellipticall way) either as an opinion, or at leastwise, as the most intelligible, and most convenient Hypothesis . . . yet in defence of Ptolemy this may be said with iustice, that there is no Astronomicall Book in the World, which may not be better spared than his [*Almagest*].

Also worth noting is what Ward said about Webster's favouring the teaching of astrology:

> But the mischiefe is, we are not given to Astrology, a sad thing, that men will not forsake the study of Arts and Languages, and give themselves up to this high and Noble Art or Science, he knows not what to call it: Nay, call it that ridiculous cheat, made up of nonsence and contradictions.

In both his statement about the teaching of Ptolemy and his ridicule of astrology, Ward conformed with the Savilian statutes, which prescribed the former and proscribed the latter.

In an appendix, Ward went after Hobbes and his prescriptions for the universities in his recently published *Leviathan*, and the tone changed. While remaining vehement in opposition, there was not the contempt that Ward showed for Webster, and Ward even remarked at the outset 'how scornefully he [Hobbes] will take it to be ranked with a Friar and an Enthusiast'. He then wrote:

> I have heard that M. Hobbs hath given out, that he hath found the solution of some Problemes, amounting to no lesse then the Quadrature of the Circle, when we shall be made happy with the sight of those his labours, I shall fall in with those that speake loudest in his praise, in the meane time I cannot dissemble my fear, that his Geometricall designe . . . could not be any thing attempted, lesse becoming such a man, as he doth apprehend himself to be.

Ward pressed his attack further, accusing Hobbes of plagiarizing his optical theory; recent research suggests that this particular accusation was unfair.[35]

The philosopher Douglas Jesseph has described Ward's appendix as 'the first shot in the war that would last the next quarter century', writing that Hobbes's fight with Wallis over the squaring of the circle was almost certainly the consequence of the goading by Ward and Wilkins.[36] Ward's own dispute with Hobbes continued in 1656 with *In Thomae Hobbii philosophiam exercitatio epistolica*, while Walter Pope added that subsequently,[37]

> before [Hobbes] would enter into an Assembly, he would enquire if Dr. Ward was there, and if he came not in, or if Dr. Ward came thither while he was there, Mr. Hobbs would immediately leave the Company. So that Dr. Ward, tho he much desird it, never had any conversation with Mr. Hobbs.

Chemistry and language

During his tenure as Savilian professor, Seth Ward's interests and efforts extended beyond astronomy, mathematics, theology, and philosophy. As a member of Wilkins's Oxford group, which he said to be about thirty in number, he reported:[38]

> Besides this Greate Clubb we have a combination of a lesser number viz: of 8 persons who have joyned together for the furnishing an elaboratory and for makeing chymicall experiments wch we doe constantly every one of us in course undertakeing by weeks to manage the worke.

This is the only occasion for which we have a record of Ward's undertaking practical and manual work for scientific research.

More important was his common cause with Wilkins to create a philosophical language. Such a language depends first upon discovering the elemental phenomena and things of the natural world and then creating one-to-one symbolic representations of each of these in the language.[39] Ward was immersed in foundational research in this area by 1650,[40] and although Wilkins is now better known than Ward for his work on the subject, in his book he thanked Ward for deepening his thinking.[41]

Ward's only published writing on the subject appeared in his blast against Webster; this subject, too, was one that Webster incorrectly accused the universities of neglecting. Ward eviscerated Webster's proposals, which would 'bring (not an advance, as this man innocently supposes, but) an elevation and uselessnesse upon Language and Grammar. For this effect is that which is pretended to by the *Universall Character*, about which he smatters so deliciously'.[42]

Ward proceeded to explain his own approach, modelled on the relatively new algebraic symbolism 'invented by Viete, advanced by Harriot, perfected by Mr Oughtred, and Des Cartes'. Although he published no more on the subject, Ward continued to engage with it many years after leaving Oxford.[43]

Later years

Within a few years of his Savilian appointment, Ward began to reveal his ambitions towards higher offices. As early as 1657, he allowed himself to be elected Principal of Jesus College, Oxford, in place of the incumbent who had been ejected. But he was still labouring under his identification with the Anglicans, and so the authorities in London refused to confirm his election.[44] Next, in September 1659, he was successfully elected President of Trinity College, Oxford, and he quickly applied his undoubted administrative talents to putting the college back in order.

In 1656 he had also become chaplain to the ejected Bishop of Exeter, at the same time accepting (and paying for) the post of precentor of Exeter, for which he was ridiculed.[45] Although it approached the limitations of the Savilian statutes, which forbade the holder of a Savilian chair from concurrently holding any ecclesiastical benefice, his precentorship was perhaps too minor to count as such, besides being dormant. By the time that Ward became President of Trinity College, the Oxford group that had formed around Wilkins was spending more time in London, and Wilkins himself had also moved on, having become Master of Trinity College, Cambridge.

Seth Ward, Bishop of Salisbury.

On 29 May 1660, Charles II arrived in London, and the interregnum was over. Ward thereupon resigned the presidency of Trinity College in favour of his ejected predecessor, and moved to London, occupying the position of a parish vicar. His precentorship of Exeter now being worth something, he took up that post also. Under the Savilian statutes, becoming a head of house required him to resign from the chair, but he did not do so until August 1660. Henceforth, Ward's career and reputation were as an eminence of the Anglican Church. In 1662 he was consecrated Bishop of Exeter, and in 1667 became Bishop of Salisbury, a grander diocese, where he remained for the rest of his life.[46]

Ward never entirely left astronomy. At a meeting on 28 November 1660, the founding members of the Royal Society of London, among them his close friends Wilkins, Wallis, and Wren, voted in a short list of additional members, including Ward.[47] He remained a participant in its meetings, and it was Ward who, late in 1671, proposed Isaac Newton for membership.[48] Newton, a generation younger, thought well of him and his library included Ward's *Idea trigonometriae demonstratae*, his treatise on comets, and his tract against Boulliau.[49] By the late 1670s Ward was still being called upon for his astronomical expertise; in 1679 he used his influence to help John Flamsteed, who was struggling to maintain his position and salary as director of the Royal Observatory at Greenwich.[50]

Few of Ward's unpublished astronomical papers survive. This is somewhat unexpected, because, even in the best of health, Ward had a poor memory for which he compensated by writing everything down.[51] Perforce, he would almost certainly have complied with the Savilian statutes' injunction that the professor should deposit his written notes.

A portrait of Christopher Wren by John Smith, based on a painting by Sir Godfrey Kneller.

Christopher Wren

In 1661 the Savilian chair passed uneventfully from Ward to the young Christopher Wren, who only a few years earlier had been Ward's student. His achievements in mathematics and astronomy were greater than those of Ward, but like Ward, Wren pursued other intellectual endeavours while occupying the chair, including the one for which he is best known.

Early years

Christopher Wren was born in 1632, the son of Christopher Wren DD. The family was profoundly Anglican, and suffered during the 1640s for the father's religious alignment.[52]

Wren's earliest education was provided by a private tutor and by his father, a man of broad interests, including curiosity about the natural world, who may have begun to instil in his son an interest in mathematics, although he was not a Copernican, favouring instead the geoheliocentric theory of Tycho Brahe.[53] William Holder, a learned man and family friend whom we met earlier advising John Greaves, was also a formative influence, and Aubrey reports that it was he who introduced Wren to arithmetic and geometry.

When Wren fell ill in 1647, his father sent him for treatment to Sir Charles Scarburgh. Wren remained with him for several months, during which Scarburgh was not only his physician but also a tutor, furthering his mathematical studies and encouraging Wren's existing interest in sundials (at the time a common recreation for the mathematically minded) and in anatomy.

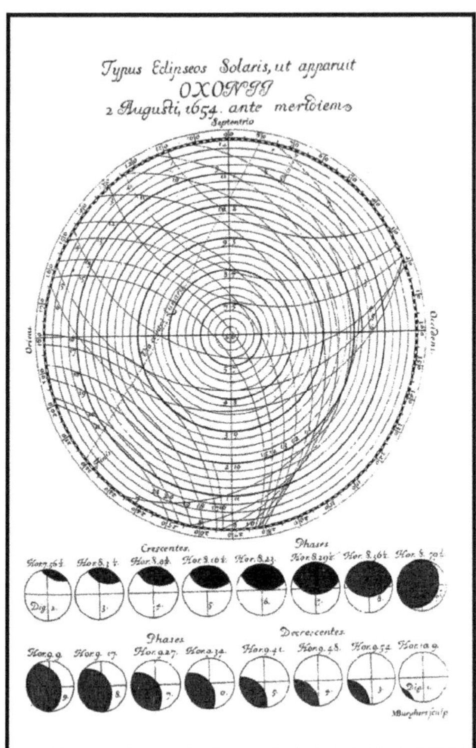

A solar eclipse was observed from Oxford on 2 August 1654 by John Wallis, Christopher Wren, and Richard Rawlinson.

He also set Wren the task of translating into Latin an English text by Oughtred on sundials and sending it to the author.

Wren is recorded as having entered Wadham College as a gentleman-commoner around 25 June 1650, although he may have arrived earlier. His son, writing long after Wren's death, singled out Seth Ward and John Wilkins as the luminaries from whom he benefitted through the Wadham connection.[54] From this time, if not earlier, Wren was identified as something of a prodigy. Entering the University was less a distinct phase of Wren's intellectual development than an acceleration along the intellectual trajectory in the sciences on which he was already propelled. By this time, in addition to his exercises in 'dialling' (sundials), he had shown a knack for constructing practical devices, including astronomical ones.[55] This continued at Wadham.

College records suggest that Wren was not paying for his rooms for much of the time that he was an undergraduate. This may be because Wilkins had taken Wren in as a lodger in the Warden's quarters, or because he was spending time at his parents' new home north of Oxford; there is circumstantial evidence for both.[56] The list of Wren's inventions and accomplishments at Oxford is very long, and only a few items bear on astronomy.[57] In 1652 William Oughtred, crediting Wren and noting that he expected great things of him, included in a new edition of his *Clavis mathematicae* Wren's translation of his earlier work on the mathematics of sundials.[58]

Wren completed his undergraduate studies in two years; although out of step with the University's statutes, this was not unprecedented, and should not necessarily be taken as an indication of exceptional precocity but rather of his cultivation by Wilkins and his circle.[59] He proceeded to an MA degree in December 1653, having just been elected a fellow of All

Souls College, where he became the college's bursar. His fellowship had previously been held by Edward Greaves (brother of John Greaves), and some faint evidence surrounding Wren's appointment suggests that, just as John had arranged for Seth Ward to succeed him in the Savilian chair, so Edward may have arranged for Wren to succeed him in his All Souls fellowship.[60] Wren's reputation at large was now such that John Evelyn, visiting All Souls in 1654, wrote in his diary 'after dinner I visited that miracle of a Youth, Mr. *Christopher Wren*, nephew to the *Bishop* of Elie'.[61] Jardine comments, however, that within the community of virtuosi he was considered a member of a team, rather than an isolated genius.[62]

When the Gresham chair of geometry became available in August 1657, Lawrence Rooke, who had left Wadham College in 1652 to become the Gresham professor of astronomy, requested a transfer from astronomy to geometry and was so appointed. With the chair of astronomy then vacant, Christopher Wren, now approaching 25 years old, was immediately appointed to the position, while retaining his fellowship at All Souls.[63] (It has been suggested that Rooke's otherwise inexplicable move was part of a scheme among Wren's patrons to secure for him the chair of astronomy.[63])

Wren's inaugural Gresham College lecture ranged widely over the history of astronomy from the Bible and classical antiquity to his own times, with few technical details but with much credit to Kepler and Galileo. He digressed on magnetism and credited William Gilbert for discovering 'the secret, and more obscure Motion of Attraction and magnetical Direction in the Earth', and thereby 'for giving Occasion to Kepler (as he himself confesses) of introducing Magneticks into the Motions of the Heavens, and consequently of building the elliptical Astronomy'.[64] To the modern reader this may seem appropriate credit, coming nearly fifty years after the publication of Kepler's *Astronomia nova* with his first two laws of planetary motion, but it was a rare mention in England (and may have been the earliest favourable one) of Kepler's underlying new celestial physics. He then went on to celebrate Kepler as 'the Compiler of another new Science, *Dioptricks*'; this relates directly to Wren's own career in astronomy, inasmuch as it concerned mainly telescopic observation. Wren's lecture then proceeded with a review of the observational discoveries from Galileo onward.

Savilian professor of astronomy

With Ward having resigned his Savilian chair in August 1660, Wren was elected as his successor on 15 February 1661. Returning to Oxford, Wren resigned his All Souls fellowship, became a fellow of Wadham College, and on 12 September of that year was awarded a Doctor of Civil Law (DCL) degree.[65] In October 1663 he moved into the college rooms formerly occupied by Ward, but in contravention of the Savilian statutes he spent only about a quarter of each year in Oxford.[66]

Wren's activities as an astronomer continued from early in his Oxford career, with his occupation of the chair marking little change in these activities or his intellectual associations, or even where he was located. In his definitive book on Wren's work in the mathematical sciences, Jim Bennett has separate chapters on 'astronomy' (meaning observational astronomy), 'cosmology' (theoretical and physical astronomy), and 'longitude'; from our perspective, as well as that of Wren's contemporaries, all three could combine under the heading of 'astronomy'. Following Bennett's convenient division, we turn now to the topics that engaged Wren while he occupied the Savilian chair.

Wren's astronomy

Christopher Wren differed from his predecessor in that, although also a mathematician, he was an astronomical observer as well as a practical man and inventor in astronomy and other sciences and fields of endeavour. Developing his skills and interests very early, he progressed rapidly from 'dialling' to cutting-edge astronomical observation as an undergraduate, under the tutelage of Sir Paul Neile, another member of Wilkins's circle. He evidently carried out his observations from both Wadham College and Neile's home, near Maidenhead.

Up to the beginning of his Savilian appointment, Wren's most important continuing work was his study of Saturn. Its peculiar appearance had been known since Galileo's first observations of it, but the early telescopes did not have the resolving power that could help observers to realize that they were seeing rings around it.

While observing with Neile in December 1657,[67] Wren formulated a hypothesis to explain its appearance and recorded this in an unpublished preliminary text, his *De corpore Saturni*. According to Wren, there was an elliptical ring around Saturn that varied in width, this width being greatest at two opposite points on an axis perpendicular to our line of sight, and least at two points on either side of the planet midway between the former pair, so that the ring touched the planet at these minimum points. Having observed that this elliptical ring varies in apparent area, as seen from the Earth, so that at times it became invisible, Wren further hypothesized that it rotated back and forth or in complete revolutions about its major axis. But in early 1659, even before its publication, Wren learned of Christiaan Huygens's explanation that circular rings surrounded the planet. Wren accepted that this hypothesis was more economical than his, and was therefore probably correct.[68]

Wren's observations of Saturn resembled those that he had made of the Moon, in that both were manifestations of a larger programme of using telescopes for systematic observation; where possible, this also included quantitative work. He carried this out in collaboration with the Wilkins group during his student years, and later with Paul Neile and Jonathan Goddard.

He also attempted to improve the capabilities of the telescope. In order to reduce both chromatic and spherical aberration, astronomers were increasing the focal lengths of objective lenses, but Wren made a unique contribution. It had long been hypothesized, originally by Kepler, that an elliptical or hyperbolic lens configuration might be preferable, and Wren was the first to propose what might have become a grinding engine to create hyperbolical lenses, as well as to prove mathematically that the surface his engine would configure was indeed of this type. But although he demonstrated such a model, there is no evidence that it was ever used.

To say that Wren used telescopes quantitatively means that he was applying them to measurement, and there were two methods for doing so. One was to apply the telescopic micrometer to measure angular distances between objects seen through the eyepiece, while the other was to mount telescopic sights onto traditional astronomical instruments (such as the quadrant) to increase the precision of sighting. The telescopic micrometer had been invented around 1640 by the northern astronomer William Gascoigne, far away from Oxford, Cambridge, and London, and independently much later by the French astronomer Adrien Auzout. The question of priority is noteworthy, because Gascoigne's invention had been obscure up until Auzout's reinvention in the 1660s, and yet knowledge of the micrometer had reached Wren and his associates in the 1650s, possibly through William Oughtred, with whom Gascoigne corresponded, and thence to Ward; indeed, it was partly documentation of Wren's use of a micrometer that

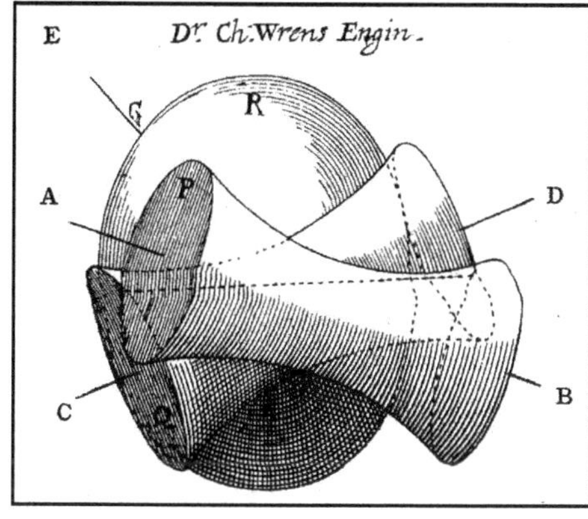

Wren's mathematical and practical insights can be seen in the design of his engine for grinding hyperbolical lenses.

convinced Auzout of Gascoigne's priority of invention.[69] As for telescopic sights, it is recorded that Wren gave Oxford a standing brass quadrant with a radius of about 26 inches, fitted with telescopic sights.[70]

In order to prove the Earth's revolution around the Sun, both Wren and Hooke attempted to apply telescopic measurement, through observation of stellar parallax with a zenith telescope situated in the Monument to the Great Fire of London, on whose design the two men had collaborated. As late as February 1704, Wren was also proposing to make such observations with a zenith telescope in St Paul's Cathedral, but another century and a half would pass before stellar parallax was measured successfully, by three astronomers working independently in the 1830s.

Determining precise longitude at sea is often considered to be the greatest technological challenge of the 17th and 18th centuries, and Wren investigated several methods for doing so. Although none of his approaches was original in conception, the determination of longitude was, as Bennett notes, the most recurrent theme in Wren's work, spanning almost his entire life. While at Gresham College, Wren had investigated the possibility of using magnetic variation (the varying deviation over the earth's surface of the magnetic compass from true north) as the basis of longitude determination, notwithstanding the prior discovery that magnetic variation, even at a given location, was not constant. It was believed that if a theory of its change could be discovered, then this could allow for correcting the measurement to a value that corresponded to precise longitude.

From this, Wren seems to have proceeded to methods based on dead reckoning (measuring the rate of water flow past a ship) and on the use of a timepiece to compare local time with the known time at a standard meridian; the concepts behind these methods predated Wren, and he explored them all, with these investigations taking place just before and during his occupancy of the Savilian chair. In 1672 Wren was appointed to a commission established by the King to consider a lunar method of longitude determination.

In his researches into physical astronomy, Wren went far beyond his Oxford mentors, Ward and Wilkins. By the time that he became Savilian professor, he was relying on Kepler's *Astronomia nova* and *Dioptrice* (a key work in modern optical theory), while being fully aware of the deficiencies of Ward's approximation method for the Kepler equation. Moreover, he accepted Kepler's proposition that an attractive force based within the Sun had to underlie a true explanation of planetary motion, although in this pre-Newtonian period he

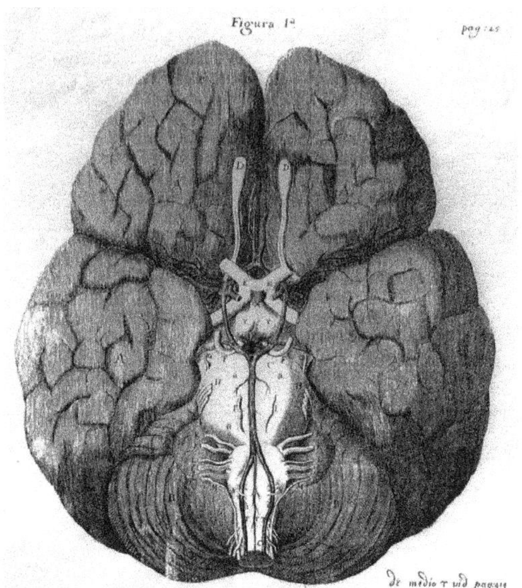

A superb illustrator, Wren produced the drawings for many of the plates in *Cerebri anatome*, a groundbreaking work on the anatomy of the brain by Thomas Willis, Oxford's Sedleian professor of natural philosophy.

could only begin to investigate how it acted. In 1659 John Wallis published Wren's solution to Kepler's equation, one that went far beyond that of his teacher (Ward) and was cited by Newton.[71]

Wren's research could also set a baseline in priority disputes. In Hooke's later struggle to ensure that Newton would grant him due credit for his earlier researches into motion under gravity, Edmond Halley (Newton's ally) solicited testimony from Wren concerning his own earlier experimental work to which Hooke had been exposed, with some of it dating back to before his occupancy of the Savilian chair. In a priority dispute with Huygens over the explanation of the oscillating pendulum, Hooke seems similarly to have relied on work done by Wren around the time that he assumed the Savilian chair.

The Savilian appointment did not lessen Wren's engagement in non-astronomical activities. Ruled by a king with a dilettantish interest in the sciences and their instruments and artefacts, Wren was on call to serve in many ways. One of these was to produce a lunar globe that would give a precise representation according to the latest measurements, while another, concurrent with the first, was to produce microscopic drawings of insects 'and small living Creatures you can light upon'. He also proposed for the King the design of a compass and odometer for a royal coach.

Apart from these obligations, and in cooperation with fellow savants, Wren pursued researches in biology and physiology and into the design of barometers and thermometers. He was one of the first to administer an intravenous drug when in 1656 he injected a dog with opium dissolved in wine. In view of his well-known architectural pursuits during his tenure of the chair (see below), it is also noteworthy that he was consulted on matters of structural engineering. In particular, Seth Ward, by now Bishop of Salisbury, commissioned from him an engineering study of Salisbury Cathedral, resulting in a report that recommended urgent repairs, especially to the spire, and with enough specificity to include the designs for ironwork to effect the necessary structural support.[72]

Christopher Wren's sundial at All Souls College.

Architecture

Christopher Wren was soon being called upon for the expertise on which rests his fame. Modern historians find his transition to architecture less anomalous than did earlier biographers, in that engineering and mathematical expertise, for which he was widely recognized, had since antiquity been considered as prerequisites for architecture.[73] Indeed, Vitruvius, the classical model to whom Wren's contemporaries looked for the rules of architecture, wrote also of mechanical devices, including sundials. In the famous paraphrase of Vitruvius by Sir Henry Wotton, 'Well building hath three Conditions. Commoditie, Firmenes, and Delight',[74] with the last word denoting a mathematical aesthetic.

Wren was not the only contemporary scholar drawn to architecture. Robert Hooke worked closely with him on building design, with both contributing to the reconstruction of London after the Great Fire of 1666.[75] There was a separate instance across the Channel: when Wren was sent to Paris in 1665, specifically to learn about the building programme under way there, he spent considerable time with the micrometer co-inventor Adrien Auzout, who also numbered architecture among his research interests.[76]

The earliest record of a contribution by Wren to architecture appeared in the list that his son provided of the 'New Theories, Inventions, Experiments, and Mechanical Improvements' that Wren apparently presented at Wadham College before the advent of the Royal Society, wherein he included 'New Designs tending to Strength, Convenience, and Beauty in Building'.[77] Historians have long identified the first architectural commission to come his way as the 1661 design of a 'mole' (a closed and fortified causeway) at Tangier, which had just become an English crown possession. This contract would have granted Wren a dispensation from the obligations of his professorship, but Wren turned it down; it seems to have come Wren's way through his cousin Matthew, secretary to the Earl of Clarendon, and because of Wren's being recognized as 'one of the best Geometricians in Europe'.[78] Concurrently, Wren seems also to have been drawn

Christopher Wren's Sheldonian Theatre, designed on an ancient Roman model. The expanse of the flat ceiling, probably based on earlier calculations and engineering models by John Wallis, caused a sensation at the time.

into plans to rebuild the former St Paul's Cathedral in London, which stood before the Great Fire.[79]

In 1663 two new architectural commissions came Wren's way, and his career in architecture fully commenced. One was for a chapel at Pembroke College, Cambridge, and the other was for the Sheldonian Theatre in Oxford. It is not known which commission came first, but the chapel, being much smaller, was completed first, and was a gift to his old college by Wren's uncle Matthew, Bishop of Ely.[80] Other commissions followed, leading inexorably to his full-scale rebuilding of the ecclesiastical architecture of London in the aftermath of the Great Fire.

As early as 1663, all of this non-astronomical activity was distracting Wren from his duties as Savilian professor, and the authorities in Oxford soon noticed. In that year, Thomas Sprat, later the first historian of the Royal Society, wrote to Wren:[81]

> I have this Occasion of telling you some ill News. The Vice-chancellor did yesterday send for me, to inquire where the Astronomy Professor was, and the Reason of his Absence, so long after the beginning of the Term . . . he most terribly told me, that he took it very ill, you had not all this while given him any Account what hinder'd you from the Discharge of your Office.

Sprat also related how he defended Wren, pointing out all that he was doing at the request of the King.

In 1669 Wren was appointed Surveyor-General of the King's Works. One may wonder how he managed to fulfil his Savilian obligations with all those attendant upon that position, not to mention all the architectural work that he had been doing since the Great Fire. There is little record of what he was doing in Oxford during these years, but we do know that his absences from the University had become insupportable, because in the year of Wren's appointment as surveyor-general he appointed Edward Bernard as his deputy as professor, a position not envisaged by the Savilian statutes. In 1673 Wren resigned the Savilian chair altogether.

The year 1669 may be taken as marking the end of Wren's career as a mathematician-astronomer. In that career, he had seldom brought his studies to completion, and he published very little. He continued to participate in meetings of the Royal Society, and his varied interests continued to occupy his thoughts, but when they did, it was away from architecture rather than from astronomy that he was drawn.

Edward Bernard

Edward Bernard may seem an unlikely choice for the Savilian professor of astronomy, as he had not previously distinguished himself in astronomical observation, instrumentation, or theory. But he was trained in mathematics and had demonstrable abilities as a scholar of the classical and semitic languages, which were useful for the recovery of early astronomical and mathematical texts. He was well connected with both English and continental scholars, and surviving correspondence shows that he commanded considerable affection and respect among his peers.

Born in 1638, Bernard matriculated in 1655 at St John's College, Oxford, his home or base for nearly all the rest of his life. He was elected a fellow of the college in 1658 and proceeded to a BA degree in 1659. By this time, he had acquired a command of Hebrew, Arabic, and Syriac. He studied mathematics privately with John Wallis, and was appointed college reader in mathematics in 1663. He proceeded to a BD degree in 1668.[82]

It seems paradoxical that Wren, frequently absent from Oxford, deputed Bernard to fulfil his duties as Savilian professor in 1669, and that Bernard then succeeded Wren to the chair in 1673, for Bernard was also absent from Oxford for long intervals during his years as deputy and as professor. We learn next to nothing about his carrying out his obligations as a lecturer and tutor, and what little we know about his making astronomical observations is slight and unimpressive. His residing at St John's College when appointed professor, thus succeeding two Wadham fellows, shows that the Savilian chair was not yet associated with any particular college.

In December 1668 Bernard departed Oxford for Leiden, for the stated purpose of collating several Arabic manuscript translations of seven books of Apollonius's *Conics*, with a view to combining these with Apollonius codices in Oxford's Bodleian library for a published edition.[83] But he did not complete this task. Indeed, a recent historian, after discussing Bernard's 'miserable failure' to bring the Apollonius edition to completion, has summarized Bernard's scholarly career thus:[84]

> He would embark enthusiastically on a new project, only to become enmired, lose interest, and leave only a trail of undigested notes.

Although textual scholarship might now seem to lie outside the range of a mathematician or astronomer, this was not so in the 17th century; in particular, Edmond Halley, as Savilian professor of geometry, eventually brought Bernard's edition of Apollonius to fruition, carrying out most of the editing and translation himself.[85] Indeed, the continuity of manuscript studies in the Savilian chairs reaches back much earlier, for some of John Greaves's holdings of Arabic manuscripts, including a couple that are referenced in John Bainbridge's *Canicularia*, came to the Bodleian in 1698 by purchase from Bernard's estate.[86]

Perhaps because he was a Savilian professor, or more simply because of his Oxford connections, Bernard was deeply implicated in the network of natural philosophers and mathematicians in England. By early 1670 he was friendly with Isaac Barrow at Cambridge, and through

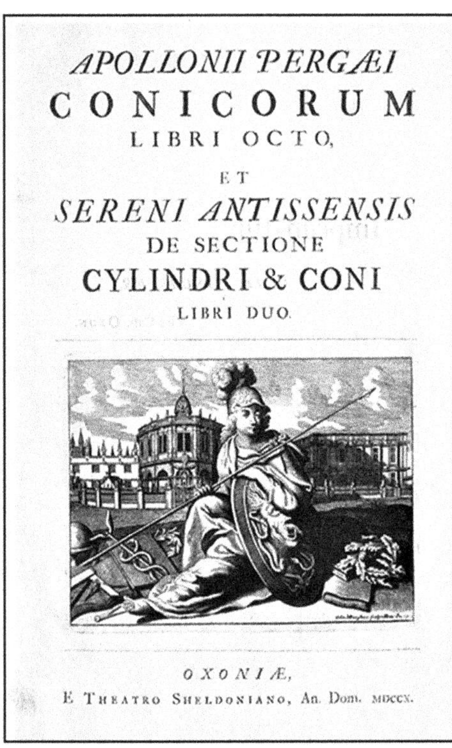

Although Edward Bernard failed to deliver a translation of Apollonius's *Conics*, an edition was later produced by Edmond Halley, John Wallis's successor as Savilian professor of geometry. Halley's frontispiece illustrates the story of Aristippus, introduced to Oxford audiences 140 years earlier by Henry Savile, and shows the Socratic philosopher, shipwrecked on the island of Rhodes, being assured of the local inhabitants' civilized nature by discovering mathematical diagrams in the sand.

him was in communication with Isaac Newton, receiving reports of his lectures when Newton at age 27 was still in the early stages of his mathematical career; indeed, a letter to Bernard in April 1671 is the earliest known record of Newton's construction of the first practicable reflecting telescope.[87]

In early 1672 Bernard wrote to Henry Oldenburg, secretary of the Royal Society, asking him to encourage Newton and Hooke to publish their work in dioptrics.[88] Bernard's associates assumed that he would be interested in recent mathematical developments. Thus, we find John Collins writing to him in the spring of 1672 with a full account of James Gregory's solution to the Kepler problem, so that this problem, which had engrossed his two immediate Savilian predecessors, was brought to his attention also.[89]

Although there remains little to show for whatever efforts Bernard made as an astronomical observer, let alone practical work in instrumentation, he nonetheless wrote specifically of the importance of such work and the deficiencies of Oxford in this regard – perhaps regretting a decline since the time of the Wilkins circle at Wadham – in a letter of 3 April 1671:[90]

> I must always affirm for the honour of my mother the University of Oxford, that if her children had the good utensils, which adorn the colleges of the Jesuits abroad, the

world would not long want good proof of their ingenuity. Patrons and tools are rather wanting than willing and fit workmen. We lack a corporation, a set of grinders of glasses, instrument makers, operators, and the like, that experiments may be well managed in this place.

When he wrote these words, Bernard was not yet a member of the Royal Society, but was elected to membership two years later, on the same day that he was fully appointed to the Savilian chair.[91]

Little testimony survives to Bernard's practical abilities as an astronomer, but we find Oldenburg writing to Christiaan Huygens in early 1676 that 'Ledit sieur Bernhard est, si ie ne me trompe fort, bon demonstrateur et Astronome', which sounds suspiciously like faint praise.[92] He maintained an enduring friendship with John Flamsteed (who referred to him as 'Brother Bernard'[93]), as evidenced in their correspondence.

In 1681 Bernard sent Flamsteed some cometary observations, presumably on the comet of late 1680, and requested that Flamsteed have a quadrant constructed for his use. Flamsteed was cordial, even as he let Bernard know what he thought of the observations. Returning them to Bernard, he wrote:[94]

> Although for the most part coarsely taken, they please me greatly because of their agreement with my own; but I foresee that they will be of little use, since more accurate measurements have been taken.

He also informed Bernard that he had placed an order for the instrument and would follow up with the craftsman.[95] Speaking even more remarkably to Bernard's abilities as an observer, Flamsteed had to explain to him that another comet, which Bernard claimed to have observed in September 1681, was not actually a comet, but Flamsteed betrayed no disdain at the mistake.[96] As for the quadrant, for all the effort that Flamsteed expended on having it constructed, Bernard made little use of it.[97]

It is evident by both word and action that Bernard was uncomfortable in the role of Savilian professor, and as early as January 1678 he was thinking of leaving the chair for 'Theologicall employment' and recommending Flamsteed as his successor, but nothing came of this. Flamsteed modestly expressed interest, but was advised by others that he could not hope to attain the position because he was from the wrong university, and that Halley would be a stronger candidate.[98]

In 1676–77 Bernard took temporary leave of his Savilian chair for a miserable year in Paris as tutor to two illegitimate sons of the King, only to return to Oxford. In 1691 he was given a living some miles south of Oxford and at last resigned the professorship. He died in 1697, having had little time to pursue scholarly work after the Savilian chair, but those final years accounted for what was perhaps his lasting legacy, a union catalogue of all manuscripts held by institutions in England and Ireland, the *Catalogi manuscriptorum Angliae et Hiberniae*.[99]

A memorial to David Gregory (1659–1708) in Oxford's University Church of St Mary the Virgin.

CHAPTER 3

The Newtonians

NATASHA BAILEY

The common characterization of Oxford as a humanistic centre has long been coupled with the idea that the university, unlike Cambridge's, contributed minimally to the emergence of mathematical physics in the aftermath of Isaac Newton's groundbreaking work. But the mathematically initiated in the early 18th century had ample reason to believe that Oxford scholars – not least David Gregory, John Caswell, and John Keill, the Savilian professors of astronomy from 1691 to 1721 – were propelling the advancement of the field. Focusing on the lives and activities of these three professors, this chapter underscores some of the ways in which they helped Newton's ideas to work their way into the core of the curricula.

Introduction

The last decade of the 17th century and the first two decades of the 18th saw three eminent mathematicians occupy the Savilian chair of astronomy: David Gregory (from 1691 to 1708), John Caswell (from 1709 to 1712), and John Keill (from 1712 to 1721). Gregory and Keill, in particular, went to great lengths to integrate Isaac Newton's groundbreaking work into their research and teaching and helped to ensure the uptake of his interventions among students for decades to come. Caswell, although less obviously significant, maintained continuity across the period.

The approaches of Gregory and Keill differed but, as this chapter suggests, they had common features. First, they each showcased the benefits of using geometrical analysis to account for the occurrence of natural phenomena, thereby helping to close the gap between mathematics and natural philosophy. Both also emphasized the ancient basis of many of Newton's ideas and, in so doing, warded off criticisms that the new (mathematical) physics was incompatible with academic natural philosophy, which had long relied upon ancient texts. Moreover, they both underscored applications of Newtonian physics in other academic domains, as well as in settings beyond the University.

Natasha Bailey, *The Newtonians*. In: *Oxford's Savilian Professors of Astronomy*. Edited by: Robin Wilson and Steven Balbus, Oxford University Press. DOI: 10.1093/oso/9780198894292.003.0003

By spotlighting the teaching and output of these Savilian professors, the chapter aims to offer a sense of their subtly varying approaches and contexts. Organized chronologically, it is structured around their biographies, which should give a sense of why, and at what stage, they were elected to the professorship, how they built upon one another's contributions, and what their tenures represented in terms of the transmission and adoption of Newton's thought.

David Gregory

Gregory, who had already showcased his mathematical acumen in his native Scotland, was an obvious choice for the professorship. Born in 1659 in Aberdeen to a family of distinguished scholars, his uncle, James Gregory, had made breakthroughs in calculus and infinite series and was the first Matheson professor of mathematics at the University of Edinburgh. David's education was typical of an ambitious young man of his social standing. He probably attended Aberdeen Grammar School before proceeding to study for his BA degree at Marischal College, Aberdeen, and for a medical degree at Leiden which he did not complete.

Soon after matriculating at Leiden in 1679, Gregory embarked on a European tour which lasted until 1681. During his travels, he became conversant with the ideas of leading Continental figures such as René Descartes, Pierre de Fermat, and Johannes Hudde; the first of these became a prime target for his later polemics against speculative natural philosophy. As for Gregory's mathematical ability, it was already on full display in his first publication, *Exercitatio geometrica de dimensione figurarum* of 1684, which built upon his uncle's work on infinite series and the geometrical properties of algebraic curves. Gregory boldly sent a copy of his work to Newton, which probably prompted the latter to write on the subject. Again following in his uncle's footsteps, he was elected as the second Matheson professor of mathematics in 1683 before even taking his MA degree – a remarkable achievement for such a young man.

Gregory's relationship with Newton developed from then on. In 1687 he acquired a copy of Newton's *Principia mathematica*, hot off the press, and it is often claimed that he was among the first professors to lecture on it anywhere. Closer examination, however, reveals that little direct material from the *Principia* made its way into Gregory's lectures.[1] He nonetheless engaged deeply with Newton's work at this point, and his commentary, *Notae in Isaaci Newtoni 'Principia'*, which he began in the late 1680s and augmented over the next twenty years, circulated widely in manuscript. Interestingly, many of Gregory's comments subtly critique Newton's ideas.[2]

Gregory was hardly in need of a job when the Savilian chair became available in 1691, for despite refusing to take an oath of allegiance to the new Presbyterian regime in Edinburgh in 1690, prominent friends had helped him to retain his position. Even so, he seized the opportunity when the 'most scholarly Dr Edward Bernard' resigned as Savilian professor and he travelled to England to promote himself as a candidate.[3]

Gregory quickly gained the backing of Newton and John Flamsteed, and was elected to the post that year. This was much to the chagrin of his fierce competitor Edmond Halley (who, however, soon became a close collaborator as the Savilian professor of geometry). According to Newton, Gregory was generally respected as 'the greatest Mathematician in Scotland & that

Oratio Inauguralis
a Davide Gregorio M: D: Astronomiæ professore Saviliano, in Auditorio Astronomico Oxoniæ, Habita vigesimo primo die Mensis Aprilis Anni 1692, quum publicam professionem auspicatus est.

Insignissime Domine Vice Cancellarie,
Reliquiq; Academiæ Procuratores honorandi.
Academici Doctissimi.

Professionis Astronomicæ munus in Academia Oxoniensi suscipientem decet in ipso principio, grata oratione Excellentissimi viri D. Henrici Savilij memoriam celebrare, qui Astronomis insomnibus illis portarum mundi vigilibus hæc otia fecit, et illustrissimis professorum Savilianorum doctoribus devinctissimum me professori, qui ea nobis quoque propria esse voluerunt.

David Gregory's inaugural lecture as Savilian professor of astronomy emphasized the value of geometry in understanding the natural world, and of the peculiarly English contribution to improving natural philosophy along these lines, fostered especially at the University of Oxford.

deservedly' and was a fine fit for the role, being 'very well skilled in Analysis & Geometry both new & Old'.[4] In a remarkably warm appraisal of Gregory's character, Newton further described him as 'prudent sober industrious modest & judicious & in Mathematiques a great Artist'. While Newton could not have known Gregory well, they met in May 1694 and Newton evidently deemed him highly capable.

An Oxonian by nature

Gregory used his inaugural lecture as Savilian professor to reveal his already impressive grasp of Newton's ideas, and to indicate the value of astronomy to natural philosophy. Natural philosophy formed the core of the BA degree at this time and was more intellectually revered than astronomy, seeking a causal rather than descriptive knowledge of nature. But the Savilians who preceded Gregory, especially Seth Ward and Christopher Wren (see Chapter 2), had already begun to break down the distinction between geometry and astronomy (as parts of the quadrivium) and, in turn, took steps to erode the gap between astronomy and natural philosophy. Gregory recognized the progress that these professors had made prior to Newton, and as he proclaimed in his inaugural lecture:[5]

> How easy and how exact and to put all in one word how geometrical, astronomy has been left to us by that most acute geometer, shall I say, or astronomer, the Right

> Revered Dr Seth [Ward], who took Kepler's work on elliptical orbits a step further by using geometry to determine the positions and species [types] of the orbits of the fixed stars.

Gregory's appointment to the Savilian chair in 1691 marked a further transformation in the teaching of astronomy and natural philosophy. His advocacy for astronomy revolved around the idea that the subject was now not only 'geometrical' but also closely related to natural philosophy, and that it was ultimately Newton who ushered in 'the age where questions that were once cosmographical are being transformed into geometrical problems'.

On a practical level, Gregory was of the view that 'Cosmical Qualities are as much easier as they are more Universall than particular ones, and the general contrivance simpler than that of Animals plants etc'. His point was that the subject of astronomy (the heavenly bodies) was simpler, and therefore easier to comprehend, than that of natural philosophy, and was thereby an excellent domain in which to develop an understanding of the laws of nature that could then be applied in the terrestrial realm.

Gregory also noted that:

> The easier, simpler and less composite these theories are, the more they will be consonant . . . with the very machine of the world which was constructed by the supreme creator with the greatest simplicity.

Beside the fact that the regularity of the movements of celestial bodies made it possible to 'subject them all to numbers', Gregory underscored that simpler phenomena are more perfect, and thereby reflect the ultimate simplicity of God. Both factors, he supposed, made astronomy an ideal gateway subject for future students of divinity who would have been expected to understand the operations of nature, as well as the relationship between the Creator and Creation.

Gregory taught regularly in Oxford, with the Savilian statutes requiring him to lecture on Tuesdays, Thursdays, and Saturdays for forty weeks of the year. Although his teaching was (according to Keill) too advanced for most undergraduates, Keill was among those who relished the chance to study under him.[6] Slightly later, John Evelyn, grandson of the renowned diarist, noted that while 'Keill was teaching physics, Professor Gregory was giving private courses in mathematics, which he restricted to ten or fifteen students, and offered to examine them weekly if they desired'. Indeed, he was keen to make himself available to sufficiently capable students wishing to learn mathematics, promising that:[7]

> If any number of schollars desire him to explain to them the Elements [of Euclid], or any other of the Mathematicall Sciences if they are alleady acquainted with the Elements, he will allow that company such a time as they among themselves shall agree upon, not less than an hour a day for three days in the week; in which time he will go through the said Science, explaining every proposition and illustrating it with such examples, operations, experiments, and observations as the matter shall require, until all the company fully apprehend and understand it.

Evidently, Gregory's more advanced courses were popular, while we also learn that he introduced a special examination, separate from the final oral disputation, for those who studied under him.

David Gregory's proposed mathematics courses

1. The first six with the Eleventh and Twelfth Books of Euclid's Elements.
2. The Plain Trigonometry, where is to be shewed the construction of naturall Sines, Tangents, and Secants, and of the tables of Logorithmes, as well of naturall numbers as of Sines, etc. The Practical Geometry, comprehending the descriptions and use of instruments and the manner of measuring heights, distances, surfaces, and solids.
3. Algebra, wherin is taught the method of resolving and constructing plain and solid problems, as well arithmeticall as geometricall; to which will be subjoyned the resolution of the indetermined arithmetical (or diophantaean) problemes.
4. Mechanicks, wherein are laid down the principles of all the sciences concerning motion; the five powers commonly so called explained, and the engines in common use reducible to those powers described.
5. Catoptricks and Dioptricks, where the effects of mirrours and glasses are shewed; the manner of Vision explained; and the machines for helping and enlarging the sight, as telescopes, microscopes, etc., described.
6. The Principles of Astronomy, containing the explication of all the most obvious Phaenomena of the Heavens from the true System of the World, and the Generation of the Circles of the Sphear thence arising. Here also is to be taught the doctrine of the Globes and their use, with the problemes of the first motion by them resolved. After this is to be demonstrated the Sphaericall Triginometry and the application thereof to Astronomy shewed in resolving the problemes of the Spere by calculation, and the construction of the tables of the first motion depending on this.
7. The Theory of the Planets, where the more recondite Astronomy is handled: that is, the Orbits of the Planets determined by observation; the tables for their motions described, and the method of constructing them taught, and the use of these tables shewed in finding the Planets' places, the Eclipses of the Luminares, etc.

Gregory's magnum opus, his *Astronomiae physicae & geometricae elementa*, was published in 1702, and promised to introduce advanced geometrical astronomy (and specifically Newton's theory of gravity and historical arguments from Newton's classical Scholia) to a wider coterie of students. This textbook was deliberately accessible, appealing to 'those who are less vers'd in the more abstruse parts of Geometry, or less concern'd about the Physical parts, [who] may pass over, and only read the Astronomy separately'.[8]

Despite Gregory's attempt to target parts of his *Astronomiae* at the less mathematically adept, it did not immediately become an undergraduate staple – indeed, unlike other contemporaneous textbooks, it is found in few college libraries. Daniel Waterland, Master of Magdalene College, Cambridge, began to recommend it in his *Advice to a Young Student*, which was first published in 1730 but was probably written in the 1710s. Yet even then he suggested it only to fourth-year students, with the warning that, along with Newton's *Opticks* and William Whiston's *Praelectiones physico-mathematicae*, it was 'more difficult to understand than any [books] before mention'd'.[9]

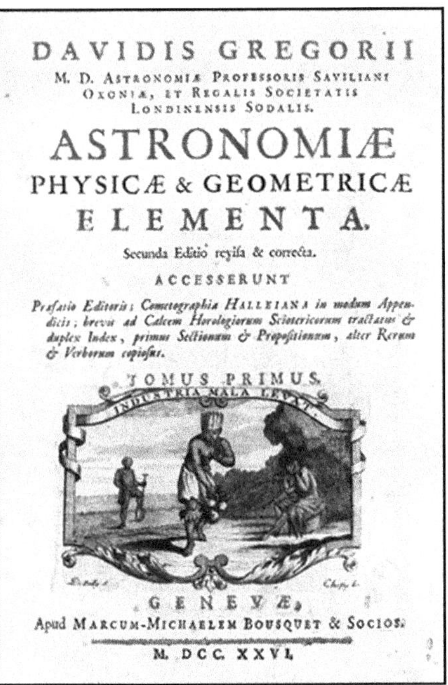

DAVIDIS GREGORII
M. D. ASTRONOMIÆ PROFESSORIS SAVILIANI
OXONIÆ, ET REGALIS SOCIETATIS
LONDINENSIS SODALIS.
ASTRONOMIÆ
PHYSICÆ & GEOMETRICÆ
ELEMENTA.
Secunda Editio revisa & correcta.
ACCESSERUNT
Præfatio Editoris; Cometographia HALLEIANA *in modum Appendicis; brevis ad Calcem Horologiorum Sciotericorum tractatus & duplex Index, primus Sectionum & Propositionum, alter Rerum & Verborum copiosus.*
TOMUS PRIMUS.
INDUSTRIA MALA LEVAT
GENEVÆ,
Apud MARCUM-MICHAELEM BOUSQUET & SOCIOS.
M. DCC. XXVI.

The frontispiece and title page of an edition of David Gregory's *Astronomia physicae & geometricae elementa*, which helped to promote the teaching of physical science in the post-Newtonian era.

At the opposite extreme, some of Newton's Continental detractors deemed Gregory's *Astronomiae* to be rather derivative. The Swiss mathematician Johann Bernoulli, for example, noted in a letter to Abraham De Moivre that 'it is only a compilation', and 'mainly of what Mr. Newton has already published in natural philosophy'.[10] Indeed, Gregory's more commonly read work in the early 18th century appears to have been his *Catoptricae et dioptricae sphaericae elementa* of 1695, which dealt with the reflection and refraction of light, and in particular with chromatic aberration (a process whereby a lens fails to focus every colour, or wavelength of light, to the same point) and its possible correction. It is, for one, mentioned several times in the widely used *Quaestiones philosophicae*, which Thomas Johnson of Magdalene College, Cambridge, compiled in the 1730s to help students to prepare for disputations.[11] Nonetheless, comments about the unoriginal nature of Gregory's *Astronomiae* should be read in the light of his pedagogical aims, representing a first attempt to build a bridge between the abstruse *Principia* and the traditional content of the BA degree.

Ancient precedents

Another way in which Gregory helped to institutionalize Newtonian physics was by stressing the classical roots of Newton's breakthroughs, with the aim of showing that:[12]

> we do still tread in the steps of the Ancients in this Physical Astronomy; inasmuch as they knew that the Celestial Bodies gravitated towards each other, and were retain'd in their Orbits by the force of Gravity; and were also apprized of the Law of this Gravity.

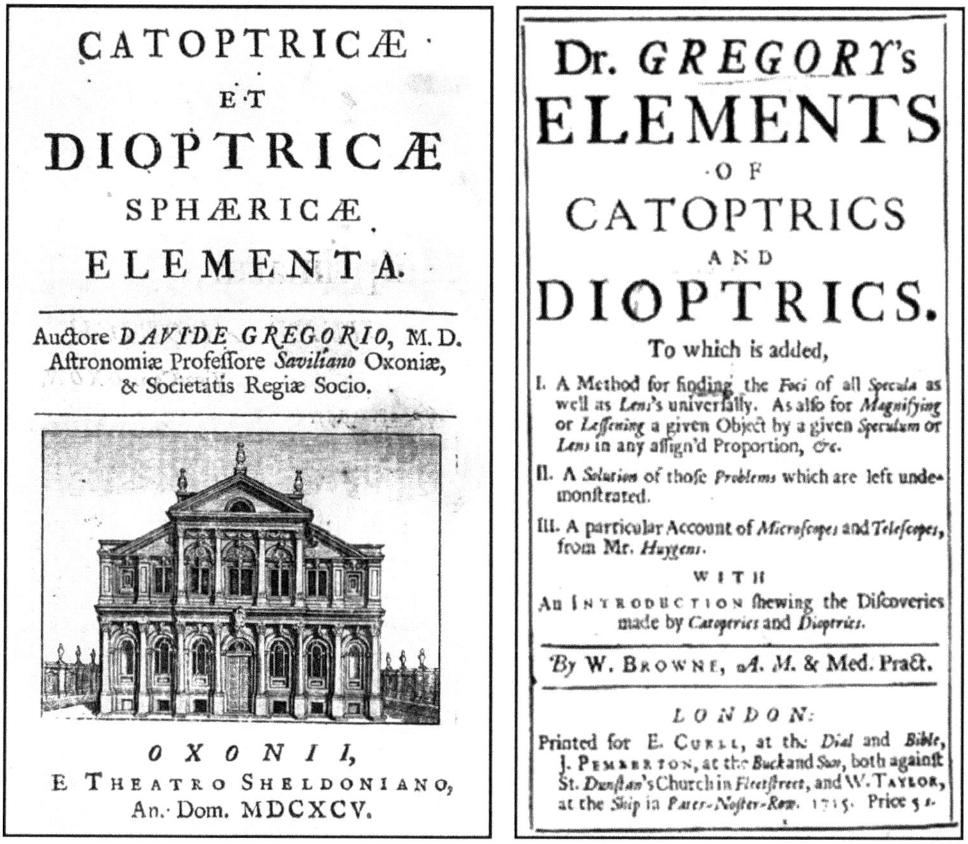

CATOPTRICÆ
ET
DIOPTRICÆ
SPHÆRICÆ
ELEMENTA.

Auctore *DAVIDE GREGORIO*, M.D.
Astronomiæ Professore *Saviliano* Oxoniæ,
& Societatis Regiæ Socio.

OXONII,
E THEATRO SHELDONIANO,
An. Dom. MDCXCV.

Dr. *GREGORY*'s
ELEMENTS
OF
CATOPTRICS
AND
DIOPTRICS.

To which is added,

I. A Method for finding the *Foci* of all *Specula* as well as *Lens*'s universally. As also for *Magnifying* or *Lessening* a given Object by a given *Speculum* or *Lens* in any assign'd Proportion, *&c.*

II. A *Solution* of those *Problems* which are left undemonstrated.

III. A particular Account of *Microscopes* and *Telescopes*, from Mr. *Huygens*.

WITH

An INTRODUCTION shewing the Discoveries made by *Catoptrics* and *Dioptrics*.

By W. BROWNE, *A. M.* & Med. Pract.

LONDON:

Printed for E. CURLL, at the *Dial* and *Bible*, J. PEMBERTON, at the *Buck* and *Sun*, both against St. *Dunstan*'s Church in *Fleetstreet*, and W. TAYLOR, at the *Ship* in *Pater-Noster-Row*. 1715. Price 5 *s*.

Two editions of David Gregory's *Catoptricae et dioptricae sphaericae elementa*.

In the dense history of astronomical philosophy that comprises most of the introduction to the *Astronomiae*, Gregory went to lengths to prove this point, stressing that there was 'nothing better approv'd of, nothing more universally entertained among the several Sects of Philosophers' than the law of universal gravitation.

Gregory's emphasis on the continuity between ancient and modern mathematical thought made a disruptive disciplinary intervention more palatable, while also allowing him to augment a new branch of undergraduate study: the history of philosophy. Aristotelian physics had long provided the backbone to undergraduate natural philosophical study, but it was being swiftly displaced in the later 17th century, and historicizing scientific and mathematical thought offered tutors a way to provide their students with a synoptic overview of natural philosophical systems without inculcating into them the doctrines of a single controversial school.[13]

On a formal level, the Savilian statutes also stipulated that:[14]

> The astronomical professor should know that it is part of his duty to interpret the whole of the *Mathematical Construction* of Ptolemy, called the *Almagest*, using the devices (*inventa*) of Copernicus, Geber and other more recent writers

– in other words, the professor should engage with ancient astronomical authors along with their modern interpretations and adaptations. A few years into his Savilian lectures, Gregory dropped Ptolemy and began to lecture on what would become his *Astronomiae*, thereby transferring from the ideas of a historical astronomer to astronomy itself – and yet the historical preface to his text indicates his desire to forge continuity between the more historical lectures that he delivered in the early years of his tenure and the more strictly mathematical ones given later.

With his emphasis on continuities between ancient and modern astronomical thought, Gregory was also forging further continuity across the Savilian professorships, for the process of mathematizing astronomy that had begun with the humanist astronomers of the 16th century had been championed in Oxford in the decades preceding his tenure. Indeed, Gregory's immediate predecessor, Edward Bernard – 'one of the most learned men of a learned age' – had a particular passion for ancient mathematics and technical literature (see Chapter 2). His study of the ancient mathematicians culminated in plans for an (unrealized) edition of all of the significant ancient and medieval mathematicians.[15] Gregory collaborated with Bernard on his project, supplying him with materials from the Bodleian Library such as an Arabic copy of Apollonius's *De sectio rationis*.[16] He also assisted Edmond Halley, the Savilian professor of geometry, with his work on ancient mathematical texts, including his Greek–Latin parallel edition of Apollonius's *Conics* of 1710.[17]

Moreover, Henry Aldrich, the Dean of Christ Church during Gregory's tenure, had become well known for backing an Oxonian research programme (which began with John Fell) that highlighted overlaps between ancient and current mathematical learning.[18] Gregory's own major contribution to this agenda was a parallel Greek–Latin edition of Euclid's complete works in 1703.[19] Bernard had planned this edition during the 1670s but had been unable to bring it to fruition due to financial difficulties, and it was left to several senior Oxford scholars to revive the project.[20] It was proposed that Gregory would 'take care of the Geometry & Reasoning' while the classical scholar John Hudson was tasked with ensuring the accuracy of the Greek and Latin.[21] But Gregory and Aldrich conspired to have Hudson's name removed from the title page and dedication, perhaps as part of an effort to strengthen the 'profiles for the two Savilians'.

In any case, the edition was a great success, and symbolized a triumph of technical scholarship that was still suitable for inclusion in the study programmes of undergraduates: Euclid's *Elements* had long been seen as 'the Text, on which the Professors in every University or Academy must read, to Young Gentlemen'.[22] Gregory's collaboration with Hudson also reflects a growing sense of the importance of adopting a 'divide-and-conquer' strategy in order to meet the required standards of increasingly specialized and competitive fields.[23]

Learning and ideals

In his inaugural lecture, Gregory had already signalled his view that an understanding of geometry was of great value in domains well beyond astronomy, asserting that:[24]

> It is by the help of geometry that all the arts necessary for improving life, such as geography, the rules of navigation, the determining of times, and others, have been carried to such an incredible pinnacle of distinction.

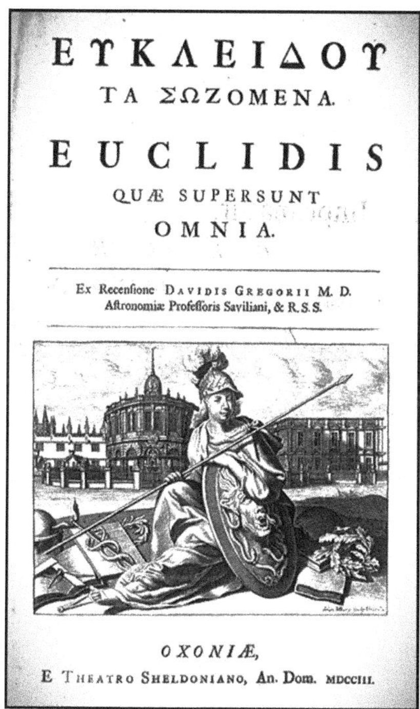
ΕΥΚΛΕΙΔΟΥ
ΤΑ ΣΩΖΟΜΕΝΑ.
EUCLIDIS
QUÆ SUPERSUNT
OMNIA.

Ex Recenſione DAVIDIS GREGORII M. D.
Aſtronomiæ Profeſſoris Saviliani, & R. S. S.

OXONIÆ,
E THEATRO SHELDONIANO, An. Dom. MDCCIII.

David Gregory's parallel Greek and Latin edition of Euclid's writings.

Not merely a posture, Gregory's own interests encompassed a number of these domains and other 'mechanical arts'. A booklist from 1693 reveals that he collected titles in medicine, cartography, naval manoeuvring, and architecture.[25] In Edinburgh, he had also delivered lectures on the kind of practical geometry that would have been required in the 'arts necessary for improving life', which were posthumously published in 1745 as *A Treatise of Practical Geometry*.[26]

In the years immediately following Gregory's appointment to the Savilian chair, practical pursuits remained a priority. He requested in 1697, for example, that the physician and collector Hans Sloane should make a copy of 'the list of all Books of Agriculture that you have in your Catalogue, and send it me. For I have a design to think on that affair somewhat', adding that 'I understand agriculture in the largest sense'.[27] All of this worked in Gregory's favour, given that the Savilian professor was required by statute 'to teach at the times as above the whole science of optics, gnomonics, geography and the precepts of navigation so far as they depend upon mathematics'.[28]

The surviving evidence from Gregory's Savilian lectures does not suggest a profound commitment to teaching the more practical subjects that depended on geometry. Nonetheless, he was keen to make mathematics useful beyond academia, motivated by a conviction that 'the Church was overstocked, and that the people encline more to mechanick Arts'.[29] To this end, he proposed to reform mathematical pedagogy in accordance with 'Foreign Colleges or Academys' and for the *Oxford Almanack* of 1703 he wrote a 'Method for Teaching Mathematicks' that focused on practical mathematical applications.[30] Broader evidence suggests that such applications ranged from the financial uses of the mathematics of infinite series to determining the most comfortable dimensions for the steps in a staircase.[31]

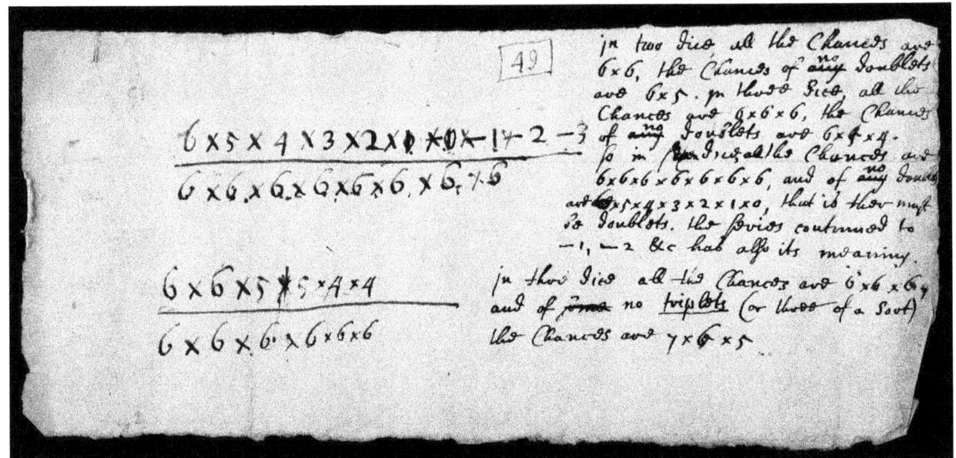
in two Dice all the Chances are 6x6, the Chances of no Doublets are 6x5. in three Dice, all the Chances are 6x6x6, the Chances of no Doublets are 6x5x4. so in six Dice all the Chances are 6x6x6x6x6x6, and of no Doublets are 6x5x4x3x2x1x0, that is there must be Doublets. the series continued to −1, −2 &c has also its meaning.
in three Dice all the Chances are 6x6x6, and of no triplets (or three of a sort) the Chances are 7x6x5.

A lighter side of David Gregory is illustrated by a handwritten note on the throwing of dice.

In 1704 Gregory set aside his teaching duties in Oxford and moved to London. He became involved in medical circles, and was made an honorary fellow of the Edinburgh College of Physicians in 1705, returning full circle to make at least a nominal use of his medical training. He also dedicated an increasing share of his time to public matters where his knowledge would prove valuable. He was, for example, appointed by the Scottish Union Commission to guide their attempt in early 1706 to calculate the 'Equivalent' (a sum that Scotland was to be paid for taking on a share of England's debt), and, through the intercession of Newton, was also made overseer of the Scottish mint in 1707, where he brought Scottish coins up to English standards.

Although Gregory did not publish further, he maintained a serious interest in scientific developments, including engaging with the debate over calculus in the *Mémoires de l'Académie royale des sciences* just before he died in Maidenhead of tuberculosis.[32] To an extent, his legacy as Savilian professor lies in his determination to weave the new mathematical physics into an ancient tapestry centred around the work of Euclid, and in the process to demonstrate that Newton's work could be similarly capable of spawning immense 'real world' changes if taught correctly.

John Caswell

John Caswell was born in 1655 and matriculated at Wadham College in 1671, where he studied under John Wallis. He graduated with a BA degree in 1674 and an MA degree in 1677, and was made a fellow of the College, where he 'taught the grounds of mathematics to young scholars'. Some years later, in around 1709, Caswell also became Vice-Principal of Hart Hall (now Hertford College), where, according to Anthony Wood, he 'carried on his faculty with great industry'.[33]

When Edward Bernard vacated the Savilian professorship in 1691, Caswell had been one of the three contenders for the post, but he did not ultimately apply due to the 'machinations of Halley', who thereby became Gregory's sole competitor.[34] When Gregory died in 1708,

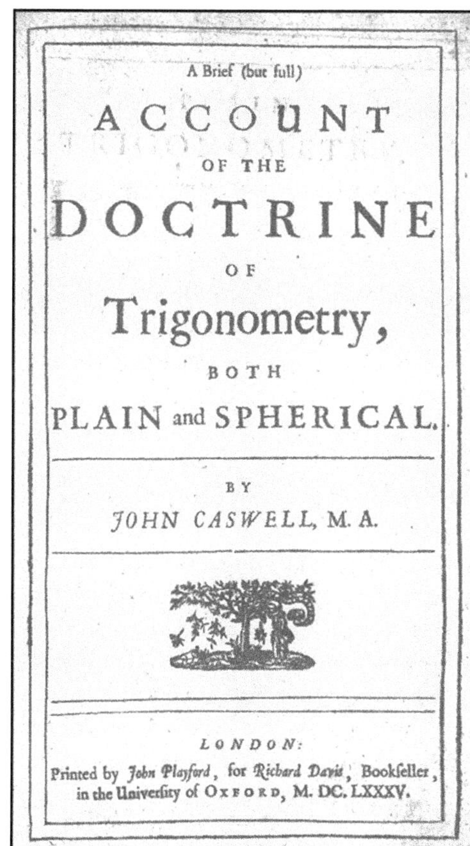

A Brief (but full)

ACCOUNT

OF THE

DOCTRINE

OF

Trigonometry,

BOTH

PLAIN and SPHERICAL.

BY

JOHN CASWELL, M. A.

LONDON:

Printed by *John Playford*, for *Richard Davis*, Bookseller, in the University of OXFORD, M. DC. LXXXV.

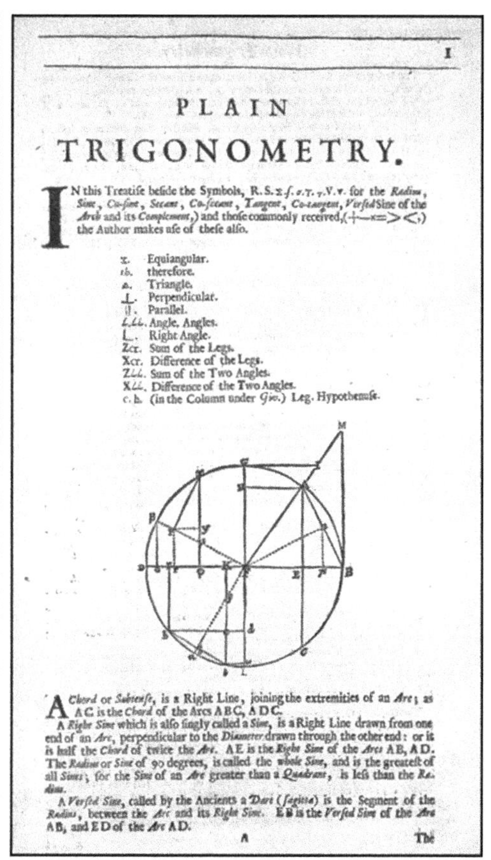

1

PLAIN

TRIGONOMETRY.

IN this Treatise beside the Symbols, R. S. s. σ. T. τ. V. v. for the *Radius*, *Sine*, *Co-sine*, *Secant*, *Co-secant*, *Tangent*, *Co-tangent*, *Versed* Sine of the *Arch* and its *Complement*,) and those commonly received, (+ − × = > <,) the Author makes use of these also.

- Equiangular.
- therefore.
- Triangle.
- Perpendicular.
- Parallel.
- Angle. Angles.
- Right Angle.
- Zcr. Sum of the Legs.
- Xcr. Difference of the Legs.
- Z∠∠. Sum of the Two Angles.
- X∠∠. Difference of the Two Angles.
- c. h. (in the Column under *Giv.*) Leg. Hypothenuse.

A *Chord* or *Subtense*, is a Right Line, joining the extremities of an *Arc*; as AC is the *Chord* of the Arcs ABC, ADC.

A *Right Sine* which is also singly called a *Sine*, is a Right Line drawn from one end of an *Arc*, perpendicular to the *Diameter* drawn through the other end: or it is half the *Chord* of twice the *Arc*. AE is the *Right Sine* of the *Arcs* AB, AD. The *Radius* or *Sine* of 90 degrees, is called the *whole Sine*, and is the greatest of all *Sines*; for the *Sine* of an *Arc* greater than a *Quadrant*, is less than the *Radius*.

A *Versed Sine*, called by the Ancients a *Dart* (*sagitta*) is the Segment of the *Radius*, between the *Arc* and its *Right Sine*. EB is the *Versed Sine* of the *Arc* AB, and ED of the *Arc* AD.

A The

John Caswell's *Doctrine of Trigonometry*, first published in 1685.

however, Caswell was ready in waiting. John Keill had also hoped to obtain the Savilian chair, and seems to have enjoyed the backing of Halley, but partly for political reasons – the Whigs had ascended to power in 1708 and Keill was known for his Tory sympathies – Caswell was elected to the chair in 1709.[35] According to the satirist Jonathan Swift, Keill's failure to obtain the position, despite being ostensibly the most accomplished candidate, confirmed that 'Party reaches even to Lines and Circles'.[36] Keill's 'vile character' is, however, also alleged to have played a role in his rejection, leading some 'sober persons' to back Caswell instead. Caswell was urged to assume the post even though it meant taking a reduction in pay of around £80 per year.[37]

Caswell is probably best known for publishing *A Brief (but full) Account of the Doctrine of Trigonometry* in 1685, whose second edition was appended to John Wallis's *A Treatise of Algebra* of the same year. Caswell's work was indeed brief, consisting of just ten pages, but nonetheless contained concise introductions to plain and spherical trigonometry, and soon became, in effect, a textbook; interestingly, it was among the reading materials of George Berkeley at Trinity College, Dublin, in the early 1700s.[38] According to Arthur Charlett, Master of University College, 'plain' trigonometry was among the mathematical subjects that was 'of most ordinary use' to students, and it accordingly became a key component of the scheme for reformed mathematics that Gregory had compiled (with Charlett's encouragement) with the 'Nobility and Gentry' in mind.[39] Not surprisingly, given the pedagogical success of this text, Caswell is listed with John Harris and Samuel Heynes as among England's principal 'Trigonometrical Writers' in Samuel Cunn's 1723 translation of *Euclid's Elements: of Geometry, from the Latin Translation of Commandine*.

Although Caswell's *Doctrine of Trigonometry* was his only attributed single-author study, he contributed to several other mathematical undertakings. Along with Gregory, for example, in 1699 he assisted John Wallis with 'A Letter of Dr Wallis to Dr Sloan, Concerning the Quadrature of the Parts of the Lunula of Hippocrates Chius, Performed by Mr John Perks', which was published in the *Philosophical Transactions*. This is another instance of the marriage of humanistic scholarship and mathematics in Oxford, as well as of collaboration between academics in the University. Caswell was also at least aware of projects such as Halley's edition of Apollonius's *Conics*, though he does not appear to have been directly involved in the editing of ancient mathematical texts.[40]

Caswell, for his part, contributed to the efforts of Gregory and Keill to shape Newton's burgeoning reputation, belittlingly describing Newton's adversary Robert Hooke as a 'good mechanic', which would have probably been taken as an insult given that the mechanical arts had remained lower on the disciplinary ladder than astronomy or natural philosophy.[41] With this said, Caswell, like Gregory, had a long-standing interest in applied mathematics and the mechanical arts, and was particularly fascinated by astronomical instruments. Between 1681 and 1684, he collaborated with John Adams on an unsuccessful 'trigonometrical' land survey of England and Wales. He also attempted to determine the height of Mount Snowdon using a barometer,[42] and in 1685 he probably authored a piece on navigation for the *Philosophical Transactions* titled 'The solutions of three chorographic problems, by a member of the Philosophical Society of Oxford'.[43]

Caswell complained in 1705 of experiencing an 'indisposition of body' and his health remained an issue, for he died prematurely from 'a much-lamented bodily infirmity' in 1712.[44] With only a three-year window of opportunity to make his mark, Caswell's tenure as Savilian professor represents more of an interval between the longer and more productive terms of Gregory and Keill than a moment of distinction. But the fact that he did not stray notably from their programmes ensured their continuity across the period, which ultimately helped to cement the new Newtonian approach.

John Keill

John Keill was born in 1671, studied under Gregory at the University of Edinburgh for his BA degree (which he obtained in 1692), and was one of Gregory's most outstanding students. In 1694, two years after Gregory obtained the Savilian professorship, Keill followed him to Oxford and matriculated at Balliol College as a senior commoner on a 'Scotch Exhibition'. Having incorporated an MA degree in 1694 he began to deliver lectures in his college rooms. Suggestive of their success, he was soon appointed lecturer in experimental philosophy at Hart Hall.[45] Here, he apparently 'mastered Newtonian philosophy' and, in the role of deputy to Sir Thomas Millington, delivered lectures and demonstrations on natural philosophy 'from a mathematical angle, which were very popular'. Among many others, John Evelyn attended them with his college friends.

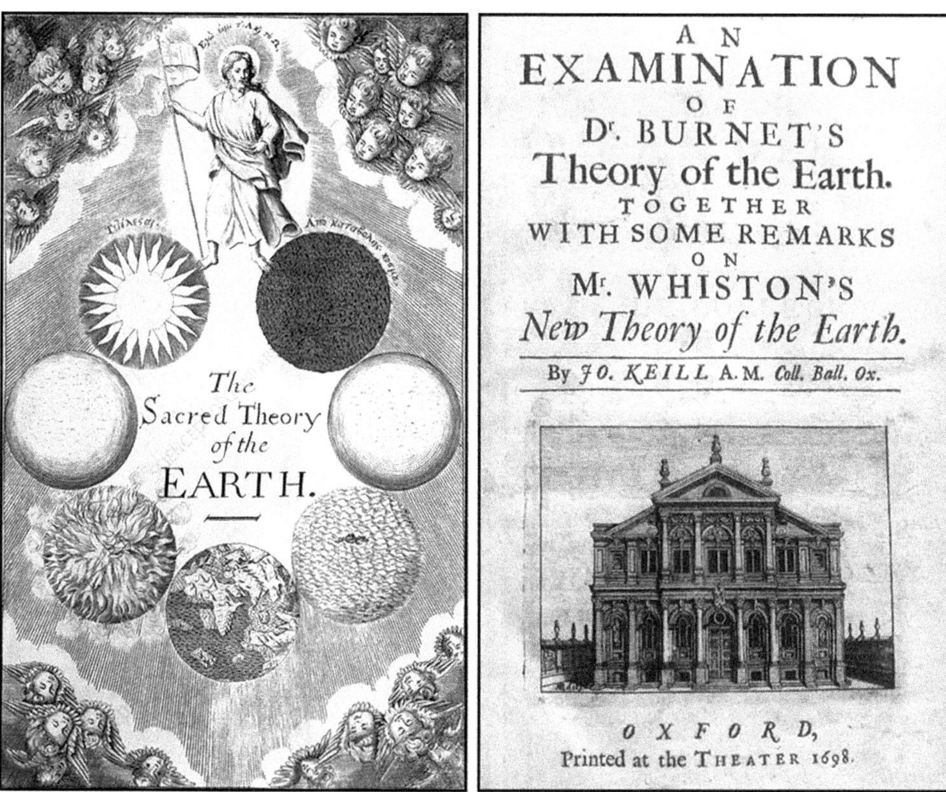

AN
EXAMINATION
OF
Dr. BURNET'S
Theory of the Earth.
TOGETHER
WITH SOME REMARKS
ON
Mr. WHISTON'S
New Theory of the Earth.

By *JO. KEILL* A.M. *Coll. Ball. Ox.*

OXFORD,
Printed at the THEATER 1698.

An English edition of Thomas Burnet's *Telluris theoria sacra* and John Keill's *Examination* of the original edition.

Keill first came to the attention of the learned world in 1698 for *An Examination of Dr. Burnet's Theory of the Earth*, a biting critique of Thomas Burnet's popular and controversial *Telluris theoria sacra* of 1681.[46] In his *Examination* Keill commended Burnet for his 'lofty and plausible stile', but attacked him for devising 'a philosophical romance', a charge often levelled at Descartes, one of Burnet's inspirations. Keill was vexed by Burnet's 'presumptuous pride' (or Cartesian confidence) in positing a single explanation for the Flood, arguing that all of Burnet's evidence 'does no way prove, that the deluge might not have been brought upon the earth by the Almighty power of God'.[47]

Keill's *Examination* was widely regarded as an exemplary demonstration of the compatibility between serious physics and a direct and literalistic reading of the Old Testament. It was swiftly being read by students, as we learn from the fact that Thomas Haywood, a fellow of St John's College, Oxford, recommended the 'answers' to Burnet (which included Keill's *Examination*) in his 1704 manuscript on 'Some Short Hints at the Method of Studying in the University'. Daniel Waterland also listed '*Burnet*'s Theory, with *Keill*'s Remarks' and '*Whiston*'s Theory, with *Keill*'s Remarks' as core reading for third-year students.[48] According to Waterland, Burnet's work (along with that of William Whiston on the same subject), and '*Keill* upon them', contain 'a great deal of curious Learning and Philosophy in them, which a Student may very much improve himself by'.[49]

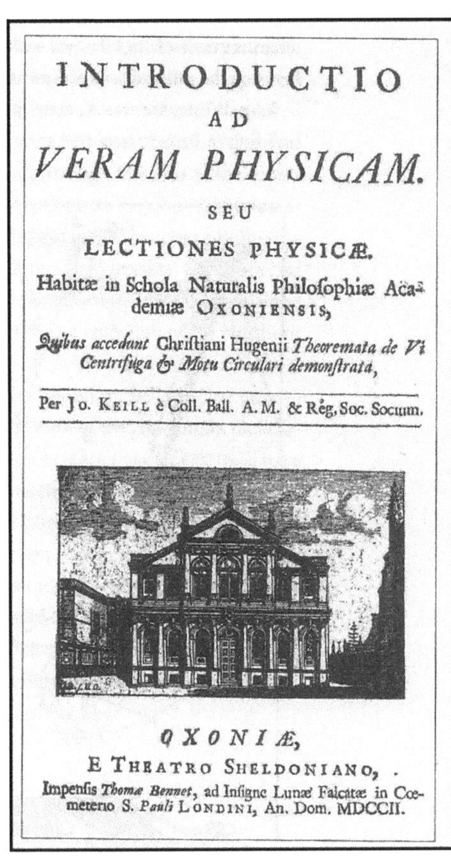

INTRODUCTIO
AD
VERAM PHYSICAM.
SEU
LECTIONES PHYSICÆ.
Habitæ in Schola Naturalis Philoſophiæ Academiæ OXONIENSIS,
Quibus accedunt Chriſtiani Hugenii *Theoremata de Vi Centrifuga & Motu Circulari demonſtrata,*
Per Jo. KEILL è Coll. Ball. A. M. & Reg. Soc. Socium.

OXONIÆ,
E THEATRO SHELDONIANO,
Impenſis *Thomæ Bennet*, ad Inſigne Lunæ Falcatæ in Cœmeterio S. *Pauli* LONDINI, An. Dom. MDCCII.

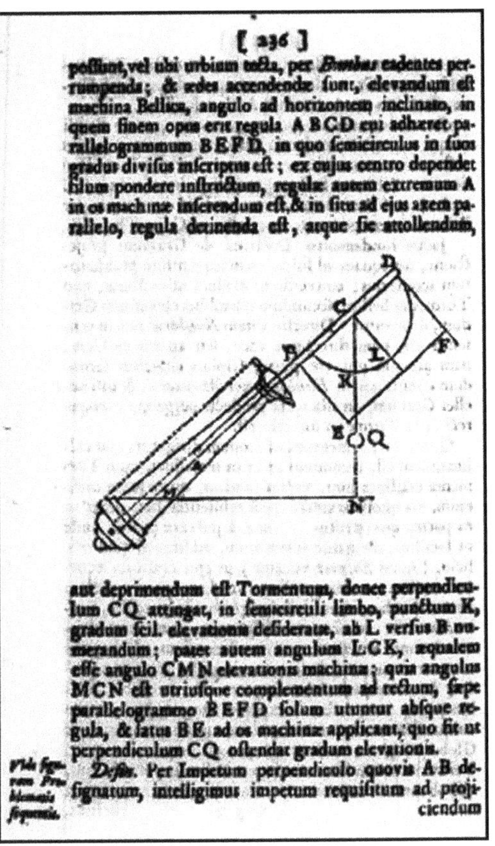

[236]

poſſint, vel ubi urbium tecta, per *Bombas* cadentes perrumpenda; & ædes accendendæ ſunt, elevandum eſt machina Bellica, angulo ad horizontem inclinato, in quem finem opus erit regula ABCD cui adhæret parallelogrammum BEFD, in quo ſemicirculus in ſuos gradus diviſus inſcriptus eſt; ex cujus centro dependet filum pondere inſtructum, regulæ autem extremum A in os machinæ inſerendum eſt, & in ſitu ad ejus axem parallelo, regula detinenda eſt, atque ſic attollendum,

aut deprimendum eſt Tormentum, donec perpendiculum CQ attingat, in ſemicirculi limbo, punctum K, gradum ſcil. elevationis deſideratæ, ab L verſus B numerandum; patet autem angulum LCK, æqualem eſſe angulo CMN elevationis machinæ; quia angulus MCN eſt utriuſque complementum ad rectum, ſæpe parallelogrammo BEFD ſolum utuntur abſque regula, & latus BE ad os machinæ applicant, quo fit ut perpendiculum CQ oſtendat gradum elevationis.

Vide figuram Problematis ſequentis. *Defin.* Per Impetum perpendiculo quovis AB deſignatum, intelligimus impetum requiſitum ad projiciendum

John Keill's *Introduction to Natural Philosophy*, published in 1702.

During the early 1700s Keill cemented his reputation as a lecturer and apologist and began to make more serious institutional inroads. In 1700 he was made a Fellow of the Royal Society, while in 1703, when his 'Scotch Exhibition' expired, he transferred to Christ Church (Gregory's college) through the influence of Aldrich. There he gave morning lectures, including some on catoptrics and dioptrics, which he seems to have delivered until at least 1707 (the date on the dictations of John Ivory, a Christ Church student).[50] As already indicated in relation to Gregory, optics was a well-studied topic at this time, perhaps because it was easier to teach through diagrams, demonstrations, and simple experiments.

Keill's Balliol lectures on natural philosophy, which began soon after his arrival in Oxford, formed the basis of his *Introductio ad veram physicam seu lectiones physicae* (Introduction to Natural Philosophy, or Philosophical Lectures) of 1702. This text confirmed his pedagogical abilities by illustrating difficult mathematical and physical concepts with experiments and demonstrations, many of which were actually thought experiments, while reducing to axioms Newton's methodological rules.[51] As the historian of mathematics Niccolò Guicciardini has noted, Keill's work actually employed mathematics quite sparingly, and relied to a greater extent on the ideas of Christiaan Huygens than those of Newton.[52]

On a methodological level, Keill's *Introductio* further demonstrated for students the advantages of taking a 'geometrical' approach to natural philosophy. As Keill promised:[53]

> We shall define things by their properties, selecting any one or several of the simplest, which by experience we know certainly belong to the things themselves; and then from these we shall deduce other properties of the same things by geometrical method.

According to this powerful statement of the 'Newtonian' methodology, geometry allows one to abstract from single instances of a physical phenomenon to form general rules, to the extent that the things in question are of the same nature. But this passage is also striking for its omission of any reference to a knowledge of causes, which had been foundational to Aristotelian–Scholastic natural philosophy and was even mentioned in Hypothesis II of the first edition of Newton's *Principia* (which Keill's point echoes in other respects). Indeed, Keill proudly went on to 'confess my ignorance frankly' of 'the intimate natures and causes of things', and in this regard followed Gregory who had shunned 'the physical causes of the Philosopher'.[54]

A growing reputation

Despite the early recognition of Keill's abilities among some of the leading scientific figures in England, he could not immediately find a satisfactory position at Oxford, partly because Caswell had taken up the Savilian chair that he coveted. A letter from Johann Bernoulli to William Burnet from August 1710 mentions that Keill was looking for employment and would accept it, even for very low wages and possibly with reference to a post in Leiden that Keill hoped to obtain.[55] Evidently, Keill did not yet have a significant reputation on the Continent either, for Bernoulli noted that he knew only of a small booklet on physics by Keill (presumably his *Introductio*) which he deemed to be rather derivative of the ideas of Newton and of those of his Continental counterparts (presumably Huygens). In response, Burnet defended Keill as a good philosopher, although he warned that he ought not to be employed in Leiden because of his libertine lifestyle.[56]

Keill's fortunes changed quite dramatically in the early 1710s, shortly before he became Savilian professor, when he helped to re-initiate the Newton–Leibniz controversy over the invention of calculus. The proceedings began in 1708 when Keill remarked in a paper in the *Philosophical Transactions* that 'Mr Newton beyond any shadow of doubt first discovered' the 'highly celebrated arithmetic of fluxions'.[57] This paper was not principally about the relative contributions of Leibniz and Newton to the discovery of calculus, but rather sought to offer 'a fluxional treatment of the inverse problem of central forces', which had recently become a pressing issue for mathematics and natural philosophy.[58] Yet Keill would surely have been aware that his comment would prove inflammatory, for it was tantamount to claiming that Leibniz had plagiarized Newton. Along with a committee of the Royal Society, and covertly guided by its president, Newton, Keill worked tirelessly to support the accusations. In particular, an anonymous collection of letters that circulated in 1713 titled the *Commercium* – in whose production Newton almost certainly had a hand – attempted to show that Leibniz had taken Newton's method of fluxions and simply changed 'the Name and Mode of Notation' to form his own differential method.[59] Leibniz in turn, and with Bernoulli's help, produced a self-defence and biting critique of Newton's position.[60] As Keill's numerous papers in Cambridge document, the debacle involved painstaking mathematical analysis and must have consumed a great deal of Keill's time.

Beyond the facts that could be used to advance the respective cases of Newton and Leibniz, there was a clear polemical backdrop to the controversy of the early 1710s that had little to do with mathematics. In 1697 Leibniz and Bernoulli had humiliated Gregory by posing a problem that was designed to test Gregory's knowledge of Leibniz's 'differential' method, and Gregory had failed their test by obtaining the solution with Newton's 'fluctional' method. As Gregory's former student and devoted follower, Keill probably had the defence of Gregory's reputation in mind when he stressed, so many years later, the priority and superiority of Newton's approach. Although Keill's efforts strained his relationship with Newton – who was reticent to enter a paper war with Leibniz because of the possibility of reputational damage – it did allow Keill to broadcast his mathematical and polemical skills.[61] Keill's abilities were finally given formal recognition in 1712 when he was appointed to the Savilian chair.

While Keill's involvement in the calculus controversy was a distraction for the first few years of his tenure, he seems to have used it to his advantage. Not only did it sharpen his mathematical abilities and keep him up to date with developments in the field, but it also provided him with a chance to teach cutting-edge material. We know, for example, that he set his more advanced undergraduates the challenge of finding solutions to problems on fluxions, since he proudly remarked in a letter from February 1715 that 'Mr Stirling an undergraduate here has likewise solved the Problem';[62] the reference here is to James Stirling, a student at Balliol College. A Cambridge student is remarked upon for having 'kept an exercise upon the 3rd section of the 1st Book of the *Principia*' and used fluxions to calculate the area of a rectilinear triangle in the late 1700s.[63] But Keill's students were clearly ahead of the game.

Keill's further publications as Savilian professor in the mid-1710s helped to cement the place of advanced geometry and Newtonian ideas within Oxford and the English universities. These included two useful books of 1715 on trigonometry and logarithms, *Trigonometriae planae & sphaericae elementa* and *Item de natura et arithmetica logarithmorum tractatus brevis*, and one from 1718 based on his astronomy lectures, *Introductio ad veram astronomiam*.[64] This last work, in particular, was a major success, while also marking an interesting methodological departure from Keill's earlier *Introductio ad veram physicam* and the writings of Gregory. Whereas Gregory had shunned causal explanations and Keill initially followed suit, Keill suggested in this later *Introductio* that it *was* desirable to study the physical causes of planetary motion, and that this same approach had enabled Newton to penetrate the mysteries of nature.

Keill's change of heart may reflect his growing recognition of what rejecting the search for causal knowledge meant for the completeness of Newton's natural philosophy and the likelihood that it was deemed a viable alternative to Aristotelianism within the universities. Indeed, at this later stage in his career, Keill's polemical tone receded more generally as he sought to develop a natural philosophy that (in the words of one commentator) represented an eclectic 'combination of "mechanical", "mathematical", "peripatetic", and "experimental"' approaches to the subject.[65]

The mysteries of nature

Just as for Gregory, Keill's interests extended well beyond mathematics, and included various mechanical arts along with classical literature, history, and 'chymistry' (the early modern

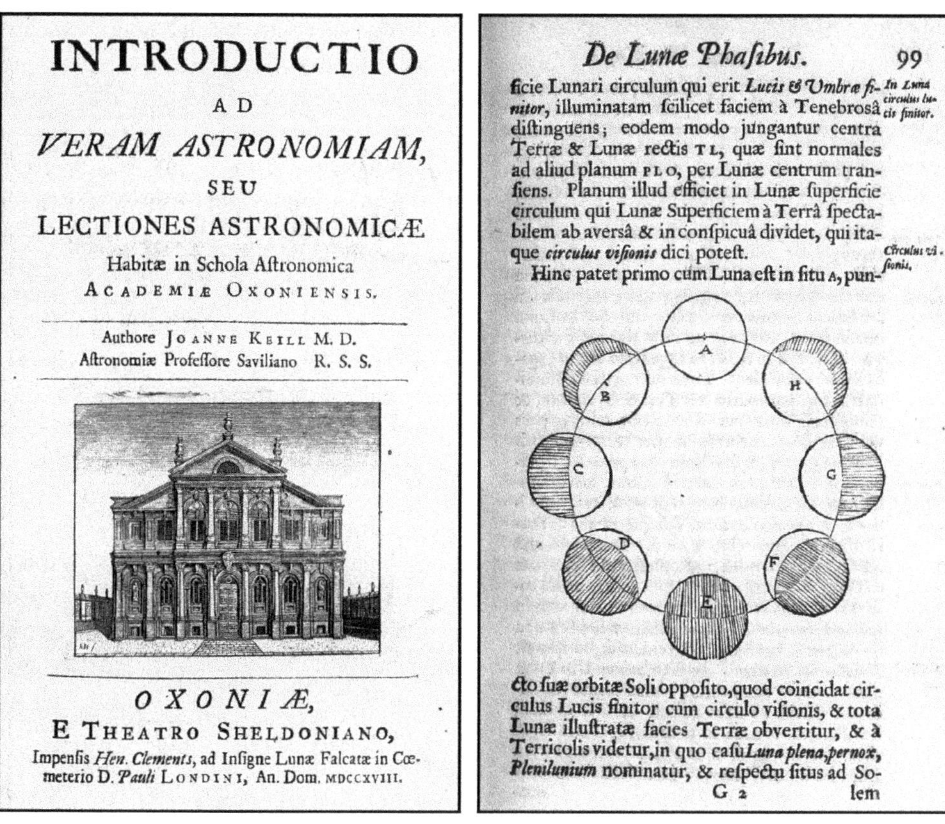

INTRODUCTIO

AD

VERAM ASTRONOMIAM,

SEU

LECTIONES ASTRONOMICÆ

Habitæ in Schola Aſtronomica

ACADEMIÆ OXONIENSIS.

Authore JOANNE KEILL M. D.
Aſtronomiæ Profeſſore Saviliano R. S. S.

OXONIÆ,

E THEATRO SHELDONIANO,

Impenſis *Hen. Clements*, ad Inſigne Lunæ Falcatæ in Cœmeterio D. *Pauli* LONDINI, An. Dom. MDCCXVIII.

De Lunæ Phaſibus. 99

ficie Lunari circulum qui erit *Lucis & Umbræ finitor*, illuminatam ſcilicet faciem à Tenebroſâ diſtinguens; eodem modo jungantur centra Terræ & Lunæ rectis TL, quæ ſint normales ad aliud planum PLO, per Lunæ centrum tranſiens. Planum illud efficiet in Lunæ ſuperficie circulum qui Lunæ Superficiem à Terrâ ſpectabilem ab averſâ & in conſpicuâ dividet, qui itaque *circulus viſionis* dici poteſt.

In Luna circulus lucis finitor.

Circulus viſionis.

Hinc patet primo cùm Luna eſt in ſitu A, puncto ſuæ orbitæ Soli oppoſito, quod coincidat circulus Lucis finitor cum circulo viſionis, & tota Lunæ illuſtratæ facies Terræ obvertitur, & à Terricolis videtur, in quo caſu *Luna plena, pernox, Plenilunium* nominatur, & reſpectu ſitus ad Solem

G 2

John Keill's *Introductio ad veram astronomiam*, published in 1718.

term for combining chemistry with alchemy). During his Savilian tenure he was especially interested in ancient astronomy and its uses, and in the draft preface to his *Introductio ad veram astronomiam* he proudly noted that 'among the Sciences there are none wch Astronomy comes behind upon the Account of its Antiquity'.[66] He added that astronomy is the 'Science wch the greatest Hero's from the beginning of the world have taken pleasure to study and improve, So that it was always esteemed as a Science fit for Kings and Emperors to employe themselves in', being what allowed monarchs 'to discover unknown and foreign Countries'.

Keill made this point more explicit with a foray into ancient mythology, remarking that astronomy had led Neptune to be celebrated as 'God of the Ocean' while allowing his son Belus, the first King of Assyria, to carry the 'inhabitants of Lybia, into Asia where he instituted Colleges of Astronomers'. According to Keill's account, the great 'Persian philosopher' Zoroaster was also skilled in astronomy, with the art being advanced thanks to the encouragement of kings in 'Africk and Syria . . . long before it was known in Greece'.[67] The nub of Keill's point was that astronomy's esteemed status was a consequence of its genuine payoff for the most ambitious actors of the past.

In his inaugural lecture as Savilian professor, apparently read early in the morning on 5 February 1713, Keill went even further in outlining the deep ancient roots of astronomy. He also offered fascinating evidence of his cyclical view of history, according to which knowledge is subject to fluctuations and does not improve linearly. This view reflected that of Francis Bacon, expressed most famously in his essay 'Of the vicissitude of things', which quite possibly inspired Keill. On a practical note, Keill warned students that knowledge could soon begin to diminish and degrade, as it had done in various historical epochs, were they not to take their studies seriously and dedicate themselves to the pursuit of truth.

On the basis of this lecture, it has been argued that Keill promoted a variant of *prisca sapientia*, or the notion that Moses and the ancient Egyptians possessed superior philosophical knowledge that the 'moderns' were only beginning to recover. In this regard, his views (like those of Gregory) have been situated in relation to the 'battle of the books' (a quarrel between defenders of ancient and modern learning respectively), with the claim being that a 'Tory' such as Keill must have come down on the side of the ancients.[68] But, as Kristine Haugen has noted, '*historia litteraria* and the study of ancient texts were enterprises that every intellectual might pursue, and ones for which many different uses could be devised'.[69] Keill (like Gregory) quite reasonably supposed that emphasizing the antiquity and power of astronomy would allow it to be upheld as the cornerstone of scientific enquiry.

Given his overarching view of history, Keill was however concerned that the amount of progress that had been recently made in astronomy (thanks largely to Newton) would mean that the main interventions that could be expected of his contemporaries would be refinements or extrapolations into more practical fields, and that knowledge could accordingly begin to stagnate. He thus pointed his students towards chymistry, botany, and comparative physiology as rich but still relatively underdeveloped fields. Keill ventured that the knowledge that one acquired about the organic realm was likely to be more speculative and not immediately useful, but this also made its pursuit an exhilarating endeavour.

Botany, in particular, had long been regarded as an inherently practical subject – a handmaiden to medicine – and so Keill's comment is perhaps suggestive of his engagement with some of the newer work that was being carried out on the nature of plant life, such as that of the physician and plant anatomist Nehemiah Grew. With palpable excitement, Keill claimed that no philosopher had yet 'discovered the figures of the small parts of matter, or the Texture Internall form and composition of the parts of the most common Plant', and that 'even in all animals and vegetable bodies, the Fountain and first Principle of life and action is unsearchable, and looks like a Mystery much beyond the reach of our own Understanding'.[70] Whereas Keill had rejected the possibility of knowing the 'natures' of bodies in his *Introductio ad veram physicam*, he now suggested that botanists and anatomists *are* able to study unseen 'natures', even if astronomers confine themselves to observable planetary 'motions' and the 'Phenomena or Appearances that arise from them'.

Gregory, in making a firm distinction between astronomical and biological studies in his inaugural lecture, had left the study of organic mysteries to others. But Keill was sufficiently enamoured by the lesser-known features of the terrestrial world (and particularly the nature of matter) to dedicate some of his time to understanding it.

Keill considered Newton's theory of matter to be one of the least satisfactory aspects of his natural philosophy, and attempted to develop a more convincing account of the kind of matter that could function within a Newtonian Universe. To this end, Keill relied on Aristotle's idea of a prime matter that 'remains the same under whatever form it may appear'. He went on to explain that because the universal qualities of bodies 'do not come from the form or modification of bodies', they must 'depend on matter itself; but because the nature of all matter is the same . . . the qualities produced by these modes will be the same in all matter'.[71] This was a clear and

106 *Of Animal Secretion.*

Corpuſcles may be ſo compounded, that the moſt ſolid and compact Particles may make up the lighteſt Corpuſcles, if the interſtices between the Particles be large, ſo that few of them may be diffuſed thro' a great Space: Such a Corpuſcle, tho' it conſiſts of Particles that are endued with a ſtrong attractive Power, may yet be ſpecifically lighter than another, which conſiſts of Particles not ſo ſolid, but cloſer together. And ſuch ſort of Corpuſcles I conceive all Salts to be, whoſe Particles of the laſt Compoſition are very ſolid, but that there are great Interſtices between thoſe Particles, into which the Water ruſhing with a force, being ſtrongly attracted, diſſolves the Texture of the Corpuſcles.

Prop. III. *If Particles of Matter attract each other with a Force, that is in a reciprocal triplicate, or a greater Proportion of their Diſtances, the*

Of Animal Secretion. 107

the Force by which a Corpuſcle is drawn to a Body made up of ſuch attractive Particles, is infinitely greater at the Contact, or extremely near it, than at any determined Diſtance from it.

Suppoſe the Sphere A H B compoſed of Particles, that attract a Corpuſcle P with a Force reciprocally proportional to the Cubes of their Diſtances. Draw the Tangent P H, and from H let fall the perpendicular H I, biſect P I in L, and

H 4 raiſe

A Newtonian physiology was developed by John Keill's brother James, whose *Account of Animal Secretion*, equipped with axioms, theorems, and attractive forces, combined Newtonian chymistry with an account of physiological phenomena.

radical statement of the matter theory that Newton would come to articulate in the second edition of his *Principia*.[72]

Keill further developed his ideas on the nature of matter and its role within a gravitational Universe after the publication of Newton's *Opticks* in 1704. He was particularly interested in Newton's remarks in Query 23 of its 1706 Latin translation, and in Newton's manuscript 'De natura acidorum' which had circulated among Keill's friends in Oxford since 1692.[73] He published his ideas on short-range chymical attractions in a 1709 paper in the *Philosophical Transactions*. Using the idea of 'special gravity' that Robert Boyle had developed to explain the inner structures of substances, Keill sought to show that matter consists of attracting particles that are arranged hierarchically, and that the forces which attract these particles act at short ranges.[74]

Keill's work on short-range attractions represented the first attempt to flesh out the chymical implications of Newton's matter theory, and laid the groundwork for his colleague, the Oxford chymist John Freind, to develop a full-fledged Newtonian chymistry in the coming years.[75] Freind divulged his views on the subject in lectures at the Ashmolean Museum, which he published in 1709 as *Praelectiones chymicae* (Chymical Lectures). In his introduction he acknowledged that his theories were heavily indebted to the work of his 'Worthy Friend' Keill.

While the product of another distinctly Oxonian research agenda, Keill's and Freind's ideas were somewhat controversial on the Continent, especially among the Leibnizians. In a review of Freind's *Praelectiones*, Christian Wolff and Leibniz argued that the theory of short-range forces reintroduced occult qualities into chymistry, and was thereby a step backwards from what they deemed to be Boyle's more precise mechanical work.[76] According to Keill's detractors, in his attempt to move Newtonian cosmology forward, he was at risk of reintroducing language and ideas that he had banished in his more 'strictly' Newtonian writings.

As regards the pedagogical integration of Keill and Freind's ideas, an anonymous student's list of physics *quaestiones* from 1722 gives us an insight into how they were combined with more traditional natural philosophical precepts. Beside each question in the list we find either 'A' (for affirm) or 'N' (for negate). The student was probably preparing for the Act (final disputations), for which he would have been randomly assigned statements to defend, and while his affirmations and negations left him defending a broadly Aristotelian cosmology he would also have concurred with certain ideas that were derived from modern authors, including Freind, such as the idea (which would have been foreign to Aristotle) that air has a *force* of elasticity.[77] The notes thus bear witness to a tendency in academic natural philosophy around this time to affirm potentially inconsistent views, in spite of (or perhaps because of) Keill's efforts to negotiate a pathway between Aristotelian and Newtonian ideas.

Having scandalously married a woman of a lower class named Mary Clements in 1717, Keill remained in Oxford and kept his professorship until his death in 1721.[78] According to Freind, he was 'a Person, who has very well deserv'd of the Philosophical World, and especially of this University'.[79] Keill had followed in Gregory's footsteps in seeking to popularize the theories of the *Principia*, tease out their implications, and deploy them within a range of mixed mathematical domains. He had also demonstrated the potential compatibility of Newton's physics with Aristotelian natural philosophy and played a role in articulating its relationship to chymistry which, much like Gregory's attempts to reveal gravity's ancient basis, was a crucial component of the broader effort to slot the new physical astronomy within the evolving institutional fold.

Conclusion

It has been argued that, following Keill's death, 'Newton became slightly marginal to Oxonian concerns', and that this reflected 'the confidence of natural philosophers within the University in their own independent experimental undertakings, the academic priority that was accorded to Literae Humaniores, and an undeniable sniffiness about the links between Newton's science and his somewhat cranky prophetic findings'.[80] While there was a brief moment of anti-Newtonian sentiment in some quarters of the University in the 1740s and 1750s – at least until the reappearance of Halley's comet in 1759 – the pedagogical changes that were initiated under Gregory, Caswell, and Keill were long-lasting.

The suggestion that Newton was marginalized in Oxford after Keill's death rests on a distinction between the 'Newtonian advocacy moment' that Gregory and Keill represented and the wider scholarly interests and pursuits of individuals in Oxford in the first half of the 18th century. But what made Gregory's and Keill's tenures so successful seems largely to have been their ability to weave their research and teaching on Newton and astronomical physics into a tapestry of other scholarly projects. In particular, their deep interests in ancient mathematical and philosophical learning allowed them to teach disruptive approaches to physics without attracting criticism from more traditional scholars who might otherwise have worried about the pace of change.

Another reason why Newton is not usually discussed in relation to post-Keillian Oxford is that rather few students are thought to have pursued advanced mathematics or physics, even by the mid-century. It is perhaps true that in Cambridge, students were more likely to be challenged in this domain – indeed, in 1750 it was said (with reference to Cambridge) that:[81]

> Mathematicks, and Natural Philosophy, are so generally, and so exactly understood, that more than twenty in every Year, of the Candidates for a Batchelor of Arts Degree, are able to demonstrate the principal Propositions in the *Principia*; and most other Books of the first Character on those subjects. Nay, several of this Number, they tell you, are no Strangers to the *higher Geometry*, and the more difficult Parts of the Mathematicks: And others, who are not of this Number, are yet well acquainted with the Experiments and *Appearances* in natural Science.

But it seems that the taking of Oxford students to the subject was simply more piecemeal and individualized – as indicated by Keill's student tackling fluxions at such an early stage.

Like their Cambridge counterparts, non-mathematically adept Oxford students benefitted greatly from the fact that, in the thirty or so years following Keill's death, there was a profusion of textbooks that helped to systematize and integrate Newton's ideas into the teaching of physics. Newton's theories were increasingly shown to be compatible with insights from domains such as moral theology, and confidence grew in the possibility of wholeheartedly replacing the Aristotelian 'system' of natural philosophy with its Newtonian successor. The major achievement of Gregory and Keill had been to begin this process – thereby showing that Newton's geometrical astronomy could form the skeleton of a new discipline whose flesh would be drawn from existing ones.

James Bradley (1692–1762), Keill's successor as Savilian professor of astronomy from 1721 and Halley's successor as Astronomer Royal from 1742.

CHAPTER 4

James Bradley

JOHN FISHER

James Bradley's early work was performed on Edmond Halley's behalf and consisted of observations of Jupiter and its four satellites. By 1720, shortly before his appointment to the Savilian chair, he had been the first to discover that the orbital motions of the three inner satellites (Io, Europa, and Ganymede) were in gravitational resonance with one other. His acute observational skills, combined with a keen mind, led directly to his discoveries of the aberration of starlight and the nutation of the Earth's axis, both of which required meticulous positional astronomy. Appointed Astronomer Royal in 1742, his mature reiterative observational practices were passed on to his assistants at the Royal Observatory. Unlike his predecessors John Flamsteed and Halley, the observational work there continued seamlessly after his death in 1762. His reform of the Observatory and its administration was his final important legacy.

Bradley's early work

The election of James Bradley as the Savilian Professor of Astronomy on 31 October 1721 was widely resented in various quarters throughout the University of Oxford. He was elected by a London-based coterie of powerful Whig politicians that included such lights as Thomas Parker, the Lord Chancellor and one-time Regent of Great Britain, and William Wake, the Archbishop of Canterbury. Much opinion within the University had favoured John Whiteside, the Chaplain of Christ Church and Keeper of the Ashmolean Museum where he gave lectures in experimental philosophy, but the matter was settled when Martin Foulkes, a future president of the Royal Society and nephew of William Wake, intervened on Bradley's behalf, reporting that his candidacy had the full support of Sir Isaac Newton.[1]

Bradley's inaugural lecture, given on 26 April 1722, was attended by Thomas Hearne, the University proctor, second librarian of the Bodleian Library, and a non-juror, who asserted that the lecture 'didn't add to his reputation'.[2] Stephen Peter Rigaud (the editor of Bradley's *Miscellaneous Works and Correspondence*) later defended Bradley, insisting that Hearne was not an acceptable and reliable judge of such matters.[3] But as a non-juror, Hearne would have

John Fisher, *James Bradley*. In: *Oxford's Savilian Professors of Astronomy*. Edited by: Robin Wilson and Steven Balbus, Oxford University Press. © Oxford University Press (2025). DOI: 10.1093/oso/9780198894292.003.0004

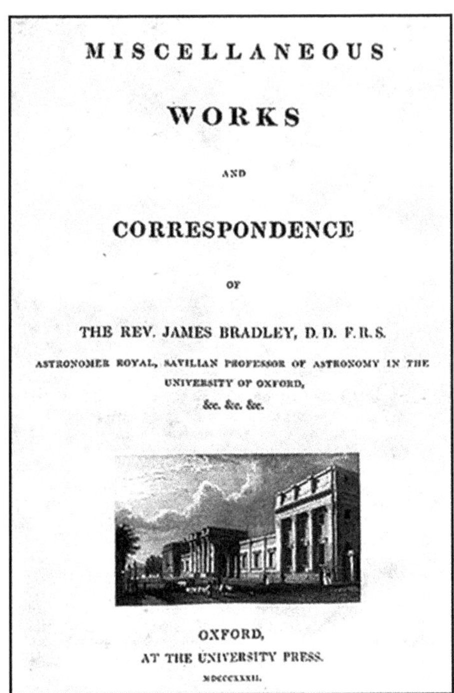
MISCELLANEOUS

WORKS

AND

CORRESPONDENCE

OF

THE REV. JAMES BRADLEY, D. D. F. R. S.

ASTRONOMER ROYAL, SAVILIAN PROFESSOR OF ASTRONOMY IN THE UNIVERSITY OF OXFORD,

&c. &c. &c.

OXFORD,

AT THE UNIVERSITY PRESS.

MDCCCXXXII.

James Bradley's *Miscellaneous Works and Correspondence*, edited in 1832 by Stephen Peter Rigaud, a later Savilian professor.

been unlikely to empathize with someone he regarded as a creature of the Whig ascendancy. Remarkably, after Whiteside's death in October 1729, when Bradley was seeking to succeed him as Keeper of the Ashmolean Museum, his *only* elector was the self-same Thomas Hearne.

Throughout his entire career, Bradley received support from all political and religious quarters, although whether this was through the integrity of his character, the power of his intellect, or simply the charm of his personality, it is difficult to ascertain. There is no question, however, that he had a remarkable ability to impress.

James Bradley was a man of modest origins, the third son of an obscure steward on a Cotswold estate, whose education at Oxford University was supported by his maternal uncle James Pound, the rector of Wanstead in Essex, with the purpose of preparing him for a career as a priest in the Church of England. Pound himself was appreciated by several within the Royal Society as one of the finest astronomical observers in England, being a pioneer of the transit instrument in England, first used by the Danish astronomer Ole Rømer. Following a breakdown of relations with John Flamsteed, the Astronomer Royal, after the unauthorized publication in 1712 of his observations as the *Historia coelestis*, Edmond Halley and Isaac Newton turned to Pound to make observations in support of their work.[4] Bradley first worked as his uncle's astronomical assistant, and then as his partner after joining him in 1711.

Halley recognized the burgeoning mathematical and observational abilities of Pound's nephew, having sufficient confidence in the young man to employ him to observe the motions of Jupiter's Galilean satellites. Halley sought to use the moons as a celestial timekeeper to determine longitude at sea,[5] but how he, as a captain in the Royal Navy, could seriously entertain the practicality of observing such motions on board a small ship at sea is difficult to credit. Bradley, however, observed and calculated the motions of the four satellites intently, producing tables for Halley, who was sufficiently impressed to propose him for the Fellowship of the

Royal Society in 1718.[6] Newton was moved to suggest that Bradley was 'the finest astronomer in Europe', showing how highly the young man was regarded.

Following the unexpected death of John Keill on 31 August 1721, the Savilian chair in astronomy became vacant and was almost immediately offered to James Pound. He was well established in Wanstead and, following the death of Flamsteed in 1719 and through the intervention of his patron Thomas Parker, had also become the rector of Burstow in Surrey. But a condition of Sir Henry Savile's bequest required the holders of the chairs in geometry and astronomy to relinquish all church sinecures, and so Pound rejected the acquisition of this prestigious chair, being unwilling to relinquish two lucrative livings for the modest stipend of the Savilian professorship. After a busy life, including several years working for the East India Company in the Indies where he narrowly escaped a massacre by mercenaries,[7] he was in a settled position with a young daughter and was about to marry Elizabeth Wymondesold, an heiress with a fortune of £10,000.

His extensive observations of the motions of Jupiter's satellites on behalf of Edmond Halley, holder of the Savilian chair in geometry and newly appointed as the second Astronomer Royal, led to Bradley becoming the first observer to comprehend that the three innermost satellites (Io, Europa, and Ganymede) were in gravitational resonance with each other. These phenomena are now commonly referred to as *Laplace resonances*, after the late 18th-century French physicist Pierre-Simon Laplace, but it was Bradley who first revealed an *exact* repetition of the cycles of mutual perturbations of these three satellites every 437 days, in accordance with Newton's universal law of gravitation.[8]

'One of the astronomy professors at Oxford'

Prestigious though the Savilian professorships were, they were unlikely to attract the interest of those with extensive church incomes, as illustrated by Pound's response to the invitation to accept the astronomy chair. His nephew James Bradley, however, was ready to relinquish his much smaller church incomes in Herefordshire and Pembrokeshire. By this time, there was no doubt that a new vision of heaven had become Bradley's chosen vocation, and on behalf of Halley he continued observing Jupiter, Saturn, and their respective satellites, while also making stellar observations on behalf of the now elderly Isaac Newton. In an age when astronomical observation strongly depended upon acute vision and a facility with observational instruments, Bradley possessed these abilities in abundance.

But it was Bradley's intellectual capacity and other attainments, including his outstanding mathematical abilities, that were most noticed by Newton and Halley. Although the focus of his lectures while Savilian professor of astronomy was slanted towards his observational skills, he also proved to be an inspired and gifted teacher of the mathematical sciences. Moreover, his observations frequently formed a basis for his teaching in the Savilian chair as when, from October to December 1723, he observed the notable comet of that year, and also shortly thereafter in early 1724 when he gave a lecture entitled *De cometa* in which he revealed his total mastery of the contents of Newton's *Principia* in a mathematical analysis and disquisition of the comet's motion.

In 1722, at the inauguration of Wanstead House, the earliest Palladian mansion in England, the Irish courtier and politician Samuel Molyneux entered into discussions with James Pound.[9] They discussed the feasibility of repeating Robert Hooke's attempts to determine annual parallax by observing the motions in declination of the star Eltanin, designated by Johann Bayer in 1603 as γ Draconis.[10] This star was chosen because, at the latitude of London,

Wanstead House, built between 1715 and 1722 on a design by the Scottish architect Colen Campbell.

it is the brightest star to pass through the meridian close to the zenith, thereby obviating the need to account for atmospheric refraction. Molyneux, who was the Prince of Wales's private secretary, was a friend of Pound's and became Bradley's patron when he obtained a half-living in the gift of the prince in Pembrokeshire to supplement the poor living in Herefordshire.

The correspondence between Samuel Molyneux and John Flamsteed, during the final years of Flamsteed's life, reveals a respectful friendship between them. Molyneux's emotive reaction to the vicious way that Newton had treated Flamsteed strongly influenced him to attempt to redress these perceived wrongs. The 'hidden script' that motivated Molyneux's repetition of Robert Hooke's earlier attempt to measure the parallax of γ Draconis can be interpreted as an attempt to cause personal distress to Newton, or even to undermine his achievement. This programme of action by Molyneux inadvertently led to Bradley's fundamental discoveries of the aberration of starlight and the nutation of the Earth's axis, laying new foundations for the high-precision science of positional astronomy, as well as for physics more generally.

Lacking the technical skills required, Molyneux sought those of his friend James Pound for his observational abilities, and of George Graham, the leading scientific instrument maker in London. But Pound's unexpected death on 16 November 1724 prompted Molyneux to invite James Bradley to take his uncle's place in an experiment at the White House at Kew Green. There can be no doubt that the wealthy and well-connected Molyneux perceived both Bradley and Graham as junior partners; indeed, he sometimes used Graham as a glorified tradesman who occasionally had to travel to Kew to make routine adjustments and repairs to the zenith sector (a suspended telescope) which he had constructed to Molyneux's commission. Molyneux referred to Bradley, holder of the respected Savilian chair in astronomy rather patronisingly, as 'one of the astronomy professors at Oxford'.[11]

Isaac Newton firmly believed that determining the annual parallax of the stars was beyond the capacity of the current available technology. He believed that the stars were at incomprehensible distances, perhaps millions of times further than the distance between the Earth and the Sun. He was also influenced by Halley, whose observational practice was based on the belief that the observation of angles finer than five arc-seconds was beyond the capacity of current instrumentation. Using James Gregory's photometric estimates,[12] Newton calculated that Sirius was at least a million times further from the Sun than the Earth, with a postulated annual parallax of less than half an arc-second. (The modern accepted value for the parallax of Sirius is 0.378 arc-seconds.) Yet for several decades, astronomers had been observing annual motions for many stars of some 30 or 40 arc-seconds per annum. These observations led to claims for the observation of annual parallax, including Hooke's assertion that the object star γ Draconis had an annual parallax of 27 or 30 arc-seconds. Flamsteed's observations of the annual motions of Polaris led to his claim of an observed annual parallax of 30 arc-seconds.[13] Both claims were based on confirmation bias, of 'seeing what you wanted to see'.

Samuel Molyneux firmly believed that the observation of annual parallax was achievable by contemporary methods. If he *had* observed an annual parallax of some 30 arc-seconds for γ Draconis, and *if* Newton's universal law of gravitation were valid, then hundreds of stars would be placed only a few thousand astronomical units away from the solar system, with all the attendant gravitational disruption to its continuing stability. Given the observed stability of the solar system, this may have been proposed as counter-evidence against the veracity of Newton's theories. This was Molyneux's hidden objective, almost certainly unknown to either of his 'junior partners', or indeed to Newton.

Bradley confirmed the southerly motion of γ Draconis on 20 December 1725, at a time when no motion due to annual parallax was anticipated, and when any expected motion would have been northward. The initial response was to suspect some unknown instrumental error, but once the southerly motion of the object star was confirmed, Bradley immediately realized that the observed motion could not be due to annual parallax. The partners were left in a quandary. A hypothesis was proposed, suggesting that the observed motion was due to an annual nutation of the Earth's axis, a word that literally means 'nodding', and indicating a small 'wobble' induced by the Sun's twice annual gravitational tugging action on the Earth's equatorial 'bulge'. Newton had proposed this twice-annual infinitesimal 'wobble' of the Earth's axis due to the Sun's gravitational interaction on the Earth's equator. Later, from 1727 to 1747, Bradley discovered a lunar induced nutation of much greater proportions, extending over a period of 18.6 years.

The problem with the design of Molyneux's experimental set-up was that it was predicated on the observation of the motions of a single star, γ Draconis. The instrument had been designed so that it could observe only those motions of the object star that were no more than a few minutes of arc from the zenith. To test the hypothesis of an annual solar induced nutation, the partners needed to find another star that passed through the meridian close to the zenith, about twelve hours in right ascension from the object star. From 3 January 1726 they sought such a control star, settling on an undistinguished light source that Bradley referred to as 'Anti-Draco' (now designated as HR2123) in the constellation of Auriga.[14] They observed it on 7 January, and by 20 February it had moved 5 arc-seconds *northwards*. The object star had been observed on 3 January and again on 13 February when it had moved 9.1 arc-seconds *southwards*. For the nutation hypothesis to be validated, the two stars should not only have moved in opposite directions (as observed), but they should also have moved through similar angular distances. After a month or so of observation, the hypothesis was declared null and void.

Molyneux's real motives were exposed by the haste of his following actions. While these observations were still being made, he launched an ill-judged and premature approach to Isaac Newton via John Conduitt, Master of the Royal Mint. Conduitt was married to Newton's niece and adopted daughter Catherine Barton, who was caring for the elderly Newton during the final months of his life. Molyneux informed Newton, via Conduitt, that he, Graham, and Bradley had discovered[15]

> a certain nutation in the earth which they could not account for and which Molyneux told me he thought destroyed entirely the Newtonian System and therefore he was under the greatest difficulty how to break it to Sir Isaac.

This approach must have been sometime during February 1726, for by the end of that month the hypothesis was recognized as untenable. Although Newton had suggested the possibility of a small annual nutation,[16] a motion of the magnitude suggested by the observations of the object and control stars was incompatible with the findings of his *Principia*.

Molyneux was doubly deluded. Newtonian natural philosophy is not a metaphysical system based on first principles, as were the natural philosophies of Aristotle or René Descartes, but was a contingent approach based on observation and experiment. According to Conduitt, Newton's own response to this premature disclosure was to assert that 'there is no arguing against facts or experiments'.[17] This null hypothesis was followed by another which proposed that the Earth was moving through a dense interplanetary medium, and the phenomenon was hypothesized as a refraction induced by the dense medium through which the Earth was moving. It could have been proposed only by someone devoid of any knowledge or comprehension of the contents of Book II of Newton's *Principia*, or from a French savant supportive of the Cartesian school; obviously this could not have been Bradley, whose knowledge of Newton's masterpiece had been mediated by his mentor Edmond Halley.

The question now arises as to why Bradley went along with such anti-Newtonian hypotheses. Molyneux was a wealthy and influential politician and courtier who had been Bradley's patron during his brief clerical career, whereas Bradley was a man of modest origins and means, who was deferential to his 'social betters' and dependent financially on his wealthy aunt Elizabeth Pound; these modest origins, as well as an inborn reticence, led to Bradley's marked deference to many of those with real power or influence. But Bradley was also a diligent and reiterative investigator who insisted that each conjecture should be tested thoroughly before any conclusions could be drawn. The fact that Bradley held the prestigious Savilian chair in astronomy seems to have meant little to Molyneux, other than to recognize Bradley's technical expertise.

Bradley made repeated visits to Kew, often finding it impossible to make observations due to inclement weather. He and Graham grew impatient, recognizing the shortcomings of the instrumental set-up at Molyneux's residence. Bradley designed an experiment that would resolve the problems associated with the counter-intuitive motions evidenced by the object and control stars at Kew, and commissioned Graham to design and construct a new zenith sector telescope capable of observing many circumpolar stars.[18]

Bradley's financial shortcomings meant that the cost was borne by his aunt Elizabeth Pound, or by Graham who was known to be a generous benefactor. At this time Graham was supporting John Harrison as he worked on his chronometers. An astute scientific worker, in 1722 Graham discovered the diurnal variation of the compass, using an instrument of his own design and construction. Like Bradley, he was a thoroughgoing empiricist, and before they made any observations or conducted any experiments, their preparations were as rigorous as the observations and experiments themselves.

At Wanstead, Bradley worked assiduously for over fourteen months, observing the motions of seventy circumpolar stars. Graham's new zenith sector was suspended at his Aunt Elizabeth's modest townhouse in August 1727, with Graham and Molyneux in attendance. There is no evidence that Molyneux ever visited Bradley again before his death on 13 April 1728, following his collapse during a debate in the House of Commons. On 2 August 1727 Molyneux had been appointed as one of the seven Lords Commissioners of the Admiralty by King George II, leaving him little time to observe γ Draconis, as revealed in the Kew Observation Book, K14. Indeed, he made only seven observations up to the end of 1727, when technical problems intervened. He then lost interest in the project, particularly following the death of Sir Isaac Newton in the same year.

George Graham FRS (*c.*1674–1751), Master of the Clockmakers Company and discoverer of the diurnal variation of the compass. His zenith sector telescope which was accurate to 0.5 arc-seconds was constructed in 1727, and with this instrument Bradley discovered the aberration of light and the nutation of the Earth's axis.

Bradley's new investigations at Wanstead were directed towards analysing the counter-intuitive motions of γ Draconis and dozens of other stars. For several decades astronomers had observed sizable motions of the 'fixed stars', either dismissing them as instrumental errors or (less commonly) making claims for the discovery of annual parallax. Flamsteed, for example, had made such a claim after observing the annual motions of Polaris for several years, writing to John Wallis, the Savilian professor of geometry, but this was repudiated by Jacques Cassini in a masterful analysis.[19] Flamsteed backtracked, accepting the French astronomer's criticisms and suggesting that he would return to his observations in order to locate the causes of the observed motions of the Pole Star. This he never did.

Bradley's Wanstead observations from August 1727 to December 1728 were rigorous and thorough. Day and night he observed the precise motions of his chosen seventy stars, making twenty or thirty precise observations of these stars on each day that he spent at Wanstead. Graham's sector was later used at the Royal Observatory, with Nevil Maskelyne, the fifth Astronomer Royal, confirming Bradley's estimation that it could dependably observe angles as small as half an arc-second.

Bradley's workload during this period was extensive. In addition to his Wanstead observations he was presenting two 45-minute term-time Latin lectures a week as Oxford's Savilian professor of astronomy. By 18th-century standards travel between Oxford and the Home Counties was convenient, with over 80% of the roads between them under the regular maintenance of turnpike trusts.[20]

During the year following August 1727, Bradley recognized that all the stars obeyed a phenomenological law. The object star γ Draconis was stationary in mid-March and mid-September, when the Earth was in quadrature with it or moving towards or away from it. From the table below, Bradley could reveal that each star reached its two stationary points, when it reversed its observed motion, while the Earth was in quadrature with each star in turn. In particular, HR2123 (the star that Bradley called 'Anti-Draco') is about twelve hours in right ascension from γ Draconis, and reaches its stationary points during March and September, one passing his meridian at 6 p.m. when the other transited his meridian at 6 a.m. The stars in the table below all followed this pattern, which was sent to Edmond Halley before it was printed in the *Philosophical Transactions*.[21]

star	*right ascension*	*declination*	*stationary periods*
α Cassiopeiae	00h 40m 40s	56° 33′ 23″	early December & early June
τ Persei	02h 54m 28s	52° 46′ 35″	late December & late June
α Persei	03h 24m 32s	49° 52′ 23″	early January & early July
α Aurigæ (Capella)	05h 16m 55s	46° 00′ 55″	early February & early August
HR2123	06h 04m 35s	51° 34′ 00″	mid-March & mid-September
γ Ursae Majoris	11h 53m 58s	53° 40′ 34″	late June & late December
ε Ursae Majoris	12h 54m 08s	55° 56′ 30″	early July & early January
η Ursae Majoris	13h 47m 37s	49° 17′ 50″	mid-July & mid-January
β Draconis	17h 30m 27s	52° 17′ 52″	early September & early March
γ Draconis	17h 56m 38s	51° 29′ 22″	mid-September & mid-March
β Cassiopeiae	00h 09m 20s	59° 10′ 09″	late November & late May

This pattern was explained in Bradley's account, addressed to Edmond Halley:[22]

> after I had continued my observations a few months, I discovered what I then apprehended to be a general law, observed by all the stars, viz. That each of them became stationary, or was farthest north or south, when they passed over my zenith at six of the clock, either in the morning or evening. I perceived likewise, that whatever situation the stars were in with respect to the cardinal points of the ecliptic, the apparent motion of every one tended the same way, when they passed my instrument about the same hour of the day or night; for they all moved southward, while they passed in the day, and northward in the night; so that each was the farthest north when it came about six of the clock in the evening, and farthest south when it came at six in the morning.

Around the time of Molyneux's death, and after more than six months of continuous observation at Wanstead, the patterns of the motions of the stars were becoming apparent.

Several years of observation and calculation of the motions of Jupiter's satellites had led Bradley to a firm confirmation of Rømer's 1676 conjecture that the velocity of light is finite, as validated by his own reiterated observations of these satellites. This conjecture, made at the Paris Observatory, contradicted the acceptance (in accordance with Cartesian theory) of the instantaneous velocity of light, then the orthodoxy in much of Europe. To Bradley, however, this was quantitative science, and clearly more than a mere conjecture. In his tables of the satellites' motions, he had to include an equation involving time. While Rømer's data was good enough for Huygens to calculate an estimate for the velocity of light of about 70% of the established modern value, the 'constant of aberration' discovered by Bradley led to a calculation that differed by less than 1% from the modern accepted value. If the velocity of light were finite, as revealed in his many observations of Jupiter's satellites, then the timings of their observed occultations would also depend on the mutual displacements of Jupiter and the Earth in their respective orbits around the Sun.

From his analyses of the motions of Jupiter's satellites and the conclusions that he drew from them, Bradley conjectured that all the observed phenomena recorded in the motions of the stars proceeded from the progressive motion of light and the Earth's annual motion in its orbit around the Sun. Having deduced the parameters of the motions of all the stars, Bradley now perceived that if light were propagated progressively in time, then the apparent location of any object would differ from when the eye is at rest or moving in a direction other than directly along the line of sight. The observed stellar displacement is, in effect, a deflection of starlight due to the motion of the Earth in its orbit around the Sun: the Earth's instantaneous velocity induces an apparent change in the direction of the velocity of light from the star observed. (An everyday analogy of this is the way that we tilt an umbrella when we walk in the rain, even when there is no wind and the rain falls vertically.) The angular scale over which these stellar deviations occur is the ratio of the Earth's orbital velocity to the speed of light, which is about 20 arc-seconds. This may be viewed as an early intimation of special relativity, in the sense that it was the first observed instance of an optical phenomenon explained by a kinematic cause.[23]

During the final weeks of 1728, Halley cajoled Bradley into writing a hurried account of 'the new discovered motion of the fixed stars', fearful that Bradley's claim might be beaten by the prior disclosure of the phenomenon by another astronomer.[24] Halley need not have been concerned. For well over sixty years, astronomers had been observing the effects of the motion now isolated by Bradley's observations at Wanstead, and many observations of what transpired to be the 'new discovered motion' were used to substantiate erroneous claims for the belief in annual parallax. These claims were the product of perceptual bias, or of 'seeing what one wants to see'.

It now seems extraordinary that so many perceptive and skilled astronomers failed to account for an annual motion of all the stars that could amount to 40 arc-seconds a year. It is a property of many profound scientific breakthroughs that in hindsight they appear so obvious. For many decades the 'aberration of starlight', as the phenomenon came to be called, remained unexplained. Astronomers had failed to identify this motion, in part because they did not trust the accuracy of their instruments and in part because any observed motions were often attributed to 'annual parallax'. The phenomenon was finally isolated and revealed by the clarity of Bradley's reasoning and the diligence of his investigations.

In summary, there were several reasons why Bradley had succeeded in identifying the new discovered motion:

1. He had immense trust in the dependability of Graham's zenith sector.
2. He was immensely knowledgeable about the parallax theory; as soon as he observed the motion of γ Draconis at Kew in December 1725, he perceived that its quantitative form could not be reduced to annual parallax.
3. His reiterative methods enabled him to determine precisely the parameters of the newly isolated motion.
4. His methodologies were founded on his belief that the laws of nature were there to be tried and tested as thoroughly as if in a court of law.

Bradley's dwelling in New College Lane. From 1672 to 1854 these two houses were occupied by successive Savilian professors, including Wallis, Gregory, Halley, Keill, and Bradley. The roof observatory was built according to Halley's instructions, and was described by Gregory as 'very convenient and indeed useful to the university, and what Sir Henry Savile did expect from them'.

The newly discovered motion of the fixed stars was soon termed as 'the aberration of light'; this terminology came with the adoption of the description by the Italian geocentrically-inclined astronomer Eustachio Manfredi of *all* observed 'motions' of the stars as 'aberrations'.[25] The term 'annual aberration' introduced by Bradley (using Manfredi's terminology) clearly distinguished the motion from 'annual parallax'. Bradley's discovery of the aberration of light also established beyond question the motion of the Earth.[26]

The laws of nature

In 1729 Bradley succeeded to the post of Lecturer (later Reader) in experimental philosophy at Oxford. This was an extra-curricular activity, and so he was able to charge fees: three guineas for a new course of twenty lectures and two guineas for a repeat. This brought him into financial self-sufficiency, but increased the demands on his time and energies at Oxford. In May 1732 he moved his main abode from Wanstead to the New College Lane accommodation that came with the Savilian bequest, and Elizabeth Pound, his widowed aunt, moved to Oxford with him. The nature of their relationship is uncertain,[27] but Georgian family law forbade marriage to a nephew. He married only after Elizabeth had died in 1740, leaving him a wealthy man.

The first two lectures of Bradley's course in experimental philosophy were a polemic directed against the teaching of the Peripatetic Schools and the contemporary rival to Newtonian empiricism that stemmed from René Descartes's variant of the mechanical philosophy, the

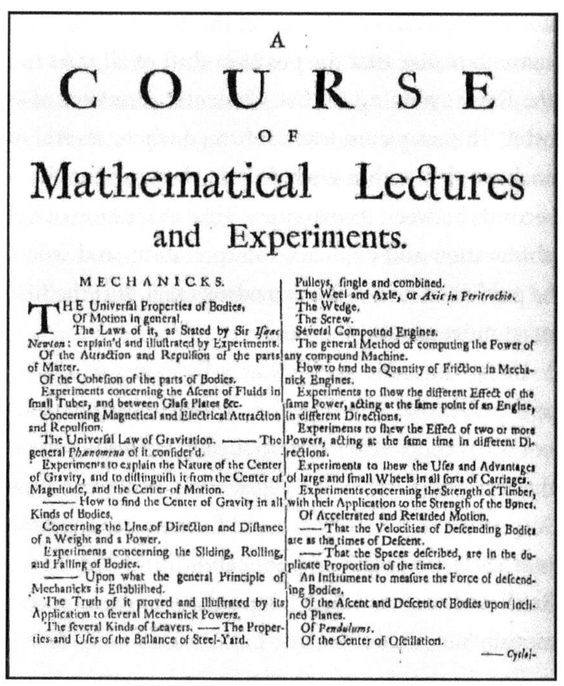

A

COURSE

OF

Mathematical Lectures

and Experiments.

MECHANICKS.

THE Universal Properties of Bodies.

Of Motion in general.

The Laws of it, as Stated by Sir *Isaac Newton*: explain'd and illustrated by Experiments.

Of the Attraction and Repulsion of the parts of Matter.

Of the Cohesion of the parts of Bodies.

Experiments concerning the Ascent of Fluids in small Tubes, and between Glass Plates &c.

Concerning Magnetical and Electrical Attraction and Repulsion.

The Universal Law of Gravitation. —— The general *Phænomena* of it consider'd.

Experiments to explain the Nature of the Center of Gravity, and to distinguish it from the Center of Magnitude, and the Center of Motion.

—— How to find the Center of Gravity in all Kinds of Bodies.

Concerning the Line of Direction and Distance of a Weight and a Power.

Experiments concerning the Sliding, Rolling, and Falling of Bodies.

—— Upon what the general Principle of Mechanicks is Established.

The Truth of it proved and Illustrated by its Application to several Mechanick Powers.

The several Kinds of Leavers. —— The Properties and Uses of the Ballance of Steel-Yard.

Pulleys, single and combined.

The Weel and Axle, or *Axis in Peritrochio*.

The Wedge.

The Screw.

Several Compound Engines.

The general Method of computing the Power of any compound Machine.

How to find the Quantity of Friction in Mechanick Engines.

Experiments to shew the different Effect of the same Power, acting at the same point of an Engine, in different Directions.

Experiments to shew the Effect of two or more Powers, acting at the same time in different Directions.

Experiments to shew the Uses and Advantages of large and small Wheels in all sorts of Carriages.

Experiments concerning the Strength of Timber, with their Application to the Strength of the Bones.

Of Accelerated and Retarded Motion.

—— That the Velocities of Descending Bodies are as the times of Descent.

—— That the Spaces described, are in the duplicate Proportion of the times.

An Instrument to measure the Force of descending Bodies.

Of the Ascent and Descent of Bodies upon inclined Planes.

Of *Pendulums*.

Of the Center of Oscillation.

—— *Cycloi-*

A syllabus for either Bradley's course of mathematics lectures in the 1740s or the similar course by his 1720s predecessor, John Whiteside.

main orthodoxy in France. Both rivals to Newtonian natural philosophy were metaphysical systems constructed on first principles. Cartesian natural philosophy was based on the fundamental ontological distinction between mind and matter where matter was identical with extension, with the Cosmos becoming a plenum. By a process of deduction, light was necessarily propagated instantaneously, as a mechanical pulse through the Cosmos. But it was the assault on Aristotle's natural philosophy that raised the greatest concern, particularly to many of those who taught classics or divinity at the University of Oxford.

Newtonian natural philosophy instigated a reversal of the traditional hierarchies of knowledge, positioning mathematics above philosophy and theology. Those who entered through the portal opened by Newton's work found a new monarch, as they saw that the world was to be perceived mathematically. The fact that Bradley was an ordained priest in the Church of England mattered little to the critics of Newtonian philosophy, several of whom believed these views to be godless and atheistic, while many conservative thinkers interpreted the whole direction of philosophy, inspired by Copernicus, in such atheistic terms. Set against Bradley's lectures in experimental philosophy, there remained a persistent core of opposition within the University.

These fears were enhanced by the growing popularity of Bradley's lectures, and by the influence that they had on the minds and sensibilities of the many students attending them. Even as matriculations declined at Oxford University, the number of students attending his lectures increased. From 1746 to 1760 there were 2788 matriculants at the University of Oxford, while this same period witnessed 1833 enrolments for Bradley's courses in experimental philosophy, with 1214 students attending one course and 619 attending two.[28] The influence of his lectures reached into many sections of English society, with one of his students being George Austen,

Shirburn Castle in Oxfordshire, home to the astronomer George Parker, 2nd Earl of Macclesfield.

the future clerical father of the novelist Jane Austen whose radicalism was subtly expressed in her works.[29] There was a radical trend in Bradley's associations, forged earlier by his uncle James Pound, as evidenced by Bradley's becoming Benjamin Hoadly's personal secretary after his ordination; Hoadly, a free thinker, was an influence on more than one of the American revolutionaries during the War of Independence.[30] The conservative reaction to Bradley's lectures must also have been tinged with an opposition to his radical associations.

After Bradley was appointed by Prime Minister Robert Walpole as the third Astronomer Royal, the University of Oxford awarded him the degree of Doctor of Divinity by diploma, a rare procedure. Later, during the autumn of 1751, he was honoured with an oration given in his presence in the Sheldonian Theatre by Dr Robert Lowth, Oxford's professor of poetry.[31] Bradley's lectures as Savilian professor of astronomy and as reader in experimental philosophy contradict a commonly expressed belief that the University of Oxford failed to embrace Newtonian natural philosophy.[32]

When the Savilian chairs in geometry and astronomy were instituted in 1619, both disciplines were regarded as branches of mixed mathematics. When Bradley was elected to the Savilian chair, astronomy was still officially conceived in applied mathematical terms, but it was not until after his death in 1762 that Oxford University acquired an observatory (see Chapter 5).[33] In the meantime he had been instrumental in the design and construction in 1739 of Lord Macclesfield's observatory at nearby Shirburn Castle.[34] Regularly used by Bradley, Nathaniel Bliss (the Savilian professor of geometry from 1742), and Thomas Hornsby (Bradley's successor as professor of astronomy), it became arguably the finest observatory in England, with an 8-foot quadrant constructed by Jonathan Sisson.

The nutation of the Earth's axis

Around the summer of 1728, Bradley took note of a motion observed within the residuals of his observations of the aberration of light. The unusual accuracy and precision of Graham's zenith sector, used to its limits by Bradley, revealed its remarkable capabilities. Initially, Bradley revisited his Kew-based hypothesis of an annual *solar* induced nutation, mentioned by Newton[35] and briefly applied when seeking an explanation for the counter-intuitive motion of γ Draconis.

Within a year Bradley began to concede that he might instead be observing a *lunar* induced nutation of the Earth's axis, which was therefore *not* an annual process, and he realized that he was committed to observing a complete retrogression of the nodes of the lunar orbit. The angle between the plane of the Earth's equator and the plane of the lunar orbit is about 5°. With one node rising and one falling, they retrogress with a period of approximately 18.6 years, creating an induced differential 'tugging' action of the Moon on the Earth's equatorial bulge. This causes the long-term precession of the equinoxes,[36] and also a miniscule nutation. Bradley observed this nutation from August 1727 to September 1747, before writing a letter to his friend and patron George Parker, 2nd Earl of Macclesfield, and a future president of the Royal Society.

After Bradley's letter was read at two meetings of the Royal Society in January 1748, he was awarded the Copley Medal by the Society's current president, Martin Foulkes. The Copley Medal, the Society's highest scientific award, was founded in 1731, too late to be awarded for Bradley's discovery of the aberration of light in 1728. Bradley's paper on nutation was important in several ways: not only was it an important observation of the perturbing effects of the Moon in conformity with Newton's universal law of gravitation, just when Cartesian natural philosophy was being widely abandoned as the primary orthodoxy in Paris, but also it enabled accurate calculations of the *annual* precession of the equinoxes for the first time since Hipparchus had first reported the phenomenon 1900 years earlier.[37]

In October 1737 Bradley first disclosed his observations of the phenomenon in a letter to the French Newtonian natural philosopher Pierre Louis Moreau de Maupertuis,[38] during the conflict between the 'Cartesians' and the 'Newtonians' at the Académie Royale des Sciences. This conflict was ultimately over what counted as evidence, and how this evidence was to be interpreted. In the Cartesian milieu, observations and experiments were undertaken to *augment* or *confirm* rational argumentation from first principles. In the Newtonian context, experiments and observations were undertaken to *establish* new knowledge. To the Newtonians the language of the new physics was mathematics, but to the Cartesians these were separate disciplines. Newton's *Principia* was perceived in Cartesian France as an exercise in geometry.

Bradley's account of aberration was rejected by most Cartesians, because his 'hypothesis' was founded on the progressive motion of light. His hypothesis of the nutation of the Earth's axis was also rejected because, following Newton, it argued for an oblate (or 'onion-shaped') Earth. Cartesian orthodoxy argued that the planets must necessarily be prolate (or 'lemon-shaped'), a belief held in the face of contrary direct observational evidence of the shapes of the giant planets Jupiter and Saturn, both glaringly obvious in any telescope as oblate spheroids.[39] The two geodesic expeditions of the 1730s, sponsored by the Académie Royal and sent to Peru and Lapland to settle the issue, were fated to fail as long as the contending parties upheld such divergent interpretations of evidence.

While Bradley was observing terrestrial nutation, he joined George Graham in a pendulum experiment to verify Newton's calculations on the shape of the Earth. Bradley and Graham were dedicated empiricists, and so those calculations needed to be tested.[40] A high-precision horological experiment was set up by the two friends in association with Colin Campbell FRS, who had an estate at Black River in Jamaica.[41] According to the Jamaican records, the thermometer varied from 15 to 20 divisions higher than those recorded in London, and this led to the conclusion that on average the clock ran some $8^1/_2$ seconds slower each day, due to the higher temperatures expanding the length of the pendulum; when compared with its performance in London the clock lost 2 minutes $6^1/_2$ seconds in Jamaica. Eliminating factors concerning the ambient air temperature in Jamaica gave a loss of 1 minute 58 seconds per day because of the diminution of gravity due to the oblate shape of the Earth. This was at variance with Newton's earlier calculations, but more importantly, it confirmed that the Earth was an oblate spheroid. Cartesian Paris ignored the results of this project.

Such was the situation during the early 1730s, when Voltaire wrote of there being an English physics and a French physics. Growing opinion in favour of Newton's contingent approach to natural knowledge was beginning to make inroads in Paris, even though there were various setbacks, such as the burning of Voltaire's *Lettres philosophique* by the public executioner in Paris in 1734 for extolling all things English, including Newtonian natural philosophy.[42]

By 1748, with the publication of Bradley's paper on the nutation and the findings of the geodetic missions, the tables had fully turned. Maupertuis, arguably the foremost supporter of Newtonian physics in the French Académie, had been elected as its director, while opinions at Court, in the salons, and in the Académie were now increasingly in favour of Newtonian natural philosophy. Much intellectual, emotional, and patriotic capital was expended defending the increasingly indefensible.

Astronomer Royal

Edmond Halley died on 14 January 1742. Because of paralysis in his right hand he had sought to stand down as the Astronomer Royal in favour of his protégé James Bradley, but this was adamantly refused by the government. With the expectation of Halley's death, many of Bradley's friends and supporters were fearful of some nonentity being appointed in Halley's place. Despite its prestige, the post of Astronomer Royal came with an annual stipend of a very modest £100 per annum, which elicited little interest. On 2 February Robert Walpole lost a division in the House of Commons over a minor issue concerning the unpopularity of his peace policy. On the very next day, he appointed Bradley as Astronomer Royal before resigning from office on 11 February.

Following his post as Astronomer Royal, Bradley was detained in Oxford, and it was not until June 1742 that he was able to travel to Greenwich to take up his new appointment. With him was John Bradley, his 14-year-old nephew and first assistant. John was to become an astute astronomer and mathematician in his own right, staying in his uncle's service until 1756 when, as a family man, he sought more lucrative employment. He eventually became Usher (second mathematical master) at the Royal Naval College in Portsmouth.

When James Bradley arrived at the Royal Observatory, what met him was a shambles. He was aware that Graham's great 8-foot quadrant was being distorted by its great weight, and the transit instrument used by Halley was so out of balance that the only way it could be used was by supporting it by hand, which was useless for such a precision instrument. The collimation mark inscribed on the wall of the Royal Park, required to adjust the line of sight of the telescope accurately, was obscured by the mature branches of an intervening tree, an indication of the amount of time that had elapsed since Halley last used the instrument.

Around 1500 observations were made between 25 July and 31 December 1742, both at Greenwich and at Shirburn Castle,[43] and throughout August Bradley made observations at twelve-hourly intervals of the first-magnitude circumpolar star Capella, above and below the pole. On 11 October George Graham added a gridiron pendulum to the clock, fixed to the brick wall facing north-west, whose design had first been developed by John Harrison.

An entry in the register for 4 December records Bradley's return from Oxford. From 16 October until that date, the entries were all in the 14-year-old John Bradley's hand;[44] his uncle had taught him many of the tacit skills, and these proved sufficient for Bradley to leave the Royal Observatory in his hands. The registers reveal that John Bradley learned how to observe the passage of a star through the field of the transit instrument while counting the vibrations of the clock, so that the declination and right ascension of a star could be fixed in a single process. This 'eye and ear' method had previously been acquired by James Bradley from his uncle James Pound when observing at his side at Wanstead.

Even with the imperfect state of the Royal Observatory's instruments, Bradley attempted several observational programmes. Not since 1689, when John Flamsteed acquired the arc that Abraham Sharp had constructed, had the Royal Observatory witnessed such levels of industry. The transit observations for 1743 alone amounted to 177 folios,[45] and on several occasions more than two hundred observations a day were recorded. The quadrant observations for the same year amounted to 148 folios,[46] involving even greater efforts than those with the transit instrument. On 8 August alone, Bradley and his nephew made 255 observations with the transit instrument and 181 observations with the quadrant. In total, around 18,000 precise observations were made in 1743.

This sustained approach to observation was characteristic of Bradley's practice, as was his commitment to accuracy in all his work. When transit observations were undertaken, the passage of stars through the wires were usually recorded to the nearest second, but Bradley commonly recorded + or – next to an entry, revealing a variation of about half an arc-second.

Reforming the Royal Observatory

Bradley regularly attempted to maintain the accuracy of the Royal Observatory's instruments, but remained dissatisfied. The comparisons that he and Lord Macclesfield made between the performances of those at Shirburn Castle and at Greenwich consistently revealed the shortcomings of the Greenwich instruments. After Bradley published his paper on nutation in January 1748 his reputation was so established that Newton's earlier assessment of him as 'the finest astronomer in Europe' was beyond dispute. In a letter sent to Bradley on 28 July 1744, the French astronomer Nicolas-Louis de Lacaille wrote:[47]

> I rejoice in the reputation you have acquired among astronomers, who have no difficulty in recognizing you as the first among them.

It was at this juncture that Bradley put forward his plans for a root-and-branch reform of the entire institution, as he sought to replace and renew the Royal Observatory's observational instruments and to acquire the resources to build a new observatory.

The old observatory with its celebrated 'Octagon room' had been designed by Christopher Wren, an astronomer of the 'old school' and an earlier Savilian professor (see Chapter 2). Wren had produced a building that enabled astronomers to locate the positions of celestial objects by triangulation, for which it was ideally constructed. But by 1675, when it opened, most positional astronomers were already locating bodies by their right ascension and declination, and the application of the quadrant and the transit instrument, when combined with the employment

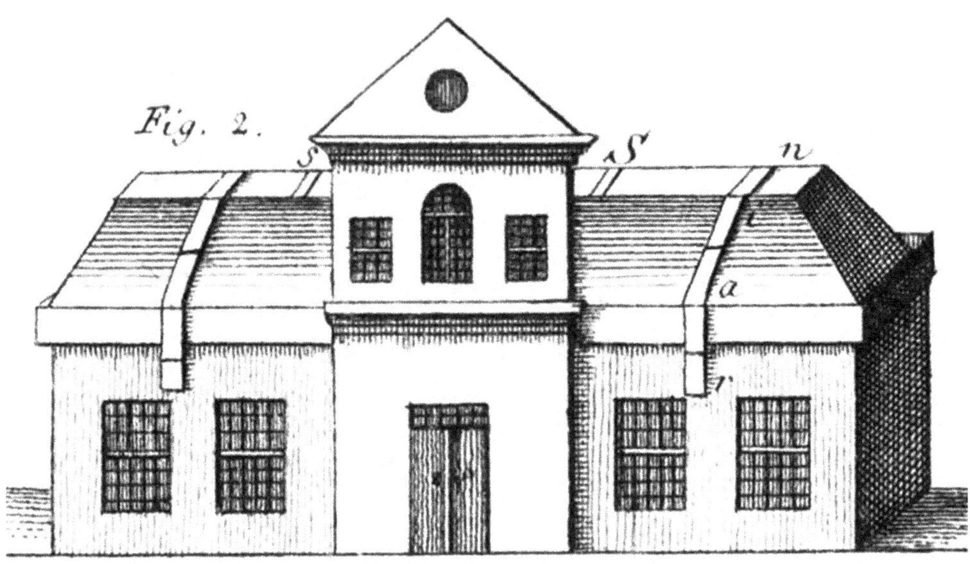

Bradley's New Observatory, pictured from the north in 1769.

of an accurate pendulum clock, furnished more dependable results for these measurements. From 1675 to 1750 successive Astronomers Royal had been forced to observe the stars from temporary outhouses, while Graham's great 8-foot quadrant was suspended on a retaining wall near the brow of the hill on which the Royal Observatory stood.

It appears that King George II had a longstanding interest in Bradley's work. His familiarity with it probably began through Bradley's associations with Samuel Molyneux, the King's private secretary from 1715 to 1727. Certainly, the announcement of Bradley's discovery of the nutation of the Earth's axis and the award of the Copley Medal in 1748 led to a rapid positive response from the King, who granted the sum of £1000 for new instruments and the repair of old ones.

Bradley began to plan a New Observatory. The temporary buildings were deemed inadequate for the required tasks now outlined by the Astronomer Royal, which included a complete survey of the stars visible from the observatory, down to the 8th or 9th magnitude. The new building, now known as the 'Transit House', is the structure through which the prime meridian passes. It may lack the charm of the edifice designed by Wren, but it was constructed for a utilitarian purpose.

The New Observatory was built around the wall that retained Graham's 8-foot quadrant, with the western end of the building housing the old quadrant and a new one made to Bradley's commission by John Bird. Cast in brass, it was lighter than Graham's original, which had been made largely of iron. During Bradley's lifetime, there was no detectable alteration in the shape of the new quadrant. Bird's instrument, modelled on that of Graham's, was also doubly divided. Both Graham and Bird marked their 8-foot quadrants with two independent scale divisions, each acting as a corrective to the other. The inner scales were divided into 90 divisions, each of one degree, which were further subdivided into twelve equal parts of five minutes of arc, and the outer scale was divided into 96 equal parts that were also subdivided. With the aid of micrometers, both scales could be used to read to the nearest arc-second. Bird's quadrant was

(*Left*) The British instrument maker John Bird (1709–76). (*Right*) Bird's 8-foot quadrant, commissioned by Bradley and modelled on Graham's earlier commission from Halley. Cast in brass, it was suspended in the New Observatory in 1750.

widely appreciated as the world's most accurate astronomical instrument, and he was quickly commissioned to make a replica for the new Russian Imperial Observatory near St Petersburg. A new transit instrument was also made by Bird in 1749, which was placed in the eastern gallery of the Transit Room, becoming the Greenwich meridian until 1816.

The 1752 calendar reform

For many years, educated opinion had sought a reform of the calendar. Great Britain was eleven days out of step with most of Europe, where the Gregorian calendar had long replaced the Julian one, and conformity with most Continental calendars had become a major objective. In England the new year had begun on Lady Day (25 March), and although Scotland had used the Julian calendar, since 1600 the year officially began on 1 January, further adding to the general confusion. Although many educated members of society saw the advantages of calendrical reform, many of the population were set against it.

During the transition from Old Style to New Style, with the adoption of the Gregorian calendar, the legal year was of only 282 days (from 25 March to 31 December 1751). The year 1752 was also a short year (of 355 days), beginning on 1 January but with the calendar advanced by eleven days, with 2 September followed by 14 September, when the Gregorian calendar was adopted. The ill-informed and credulous were led to believe that they had lost several days of their lives. After the Act of Parliament had been passed there were said to have been riots around the country (including London), with calls of 'Give us back our eleven days!' – although historians now believe this to be largely a myth.

Bradley became involved in this calendar reform, because the calculations and confirmations supporting this Act were made by him in his official role as Astronomer Royal. Ten years

(*Left*) Nathaniel Bliss (1700–64), Savilian professor of geometry (succeeding Edmond Halley) and fourth Astronomer Royal. (*Right*) 'This sure is Bliss, if Bliss on Earth there be'. A drawing of Nathaniel Bliss, scratched on a pewter flagon during dinner by the astronomer George Parker (later 2nd Earl of Macclesfield). Here, Bliss is incorrectly labelled as Oxford's Professor of Astronomy.

later, locals in Gloucestershire believed that his painful terminal illness and death were divine punishments for robbing people of eleven days of their lives!

Surveying stars from the Royal Observatory

From 1750, and up to and beyond Bradley's death on 13 July 1762, the survey of stars visible from the Royal Observatory was continued by Nathaniel Bliss, Savilian professor of geometry and Bradley's successor as Astronomer Royal. An important facet of Bradley's legacy was that he left the Royal Observatory as a viable institution, so that its work could continue unabated after his death.

Both Bradley and Halley had inherited a Royal Observatory in a state of chaos. On 10 August 1750 John Bird's new quadrant was ready for observations, and was used to survey the stars north of the Observatory, while Graham's quadrant remained in place on the other side of the retaining wall and continued to be used for observing the stars south of the Royal Observatory until it was taken down in July 1753. It was then recalibrated by Bird and repositioned to observe the stars to the north of the Observatory from August 1753, while Bird's quadrant was suspended so as to observe the stars to the south from 31 July. Bradley had transformed the Royal Observatory into the finest in Europe, and created a tradition of excellence that was inherited by his successors.

Some authorities have criticised Bradley for his 'failure' to reduce his great series of more than 60,000 unprecedented observations to something manageable. Francis Baily,[48] Flamsteed's earliest biographer, did not appreciate that the reduction of most of Flamsteed's observations was impossible after Bradley had discovered the aberration of light. What Bradley left to posterity

[9]

A CORRECT TABLE
OF THE
Longitude and Latitude of the principal Zodiacal Stars proper to take the Moon's Diſtance from, for finding the Longitude at Sea.
Deduced from Dr. Bradley's Obſervations.

Beginning of 1767.	Magnitud.	Longitude. S. ° ′ ″	Latitude. ° ′ ″
γ Pegaſi	2	0. 5. 54. 38	12. 35. 35 N
* α Arietis	2	1. 4. 24. 20	9. 57. 30 N
α Ceti	2	1. 11. 3. 56	12. 36. 16 S
* Aldebaran	1	2. 6. 32. 3	5. 29. 2 S
β Tauri	2	2. 19. 19. 19	5. 21. 59 N
α Orionis	1	2. 25. 30. 5	16. 3. 31 S
* Pollux	1. 2	3. 20. 0. 16	6. 40. 5 N
Procyon	1	3. 22. 34. 29	15. 58. 8 S
* Regulus	1	4. 26. 35. 31	0. 27. 27 N
β Leonis	2	5. 18. 23. 9	12. 17. 8 N
* Spica Virginis	1	6. 20. 35. 31	2. 2. 11 S
α Libræ	2	7. 11. 50. 11	0. 21. 48 N
β Libræ	2	7. 16. 7. 23	8. 31. 32 N
* Antares	1	8. 6. 30. 40	4. 32. 17 S
σ Sagittarii	2. 3	9. 9. 7. 59	3. 24. 55 S
* α Aquilæ	1	9. 28. 29. 13	29. 18. 36 N
* β Capricorni	3	10. 0. 47. 37	4. 36. 46 N
* Fomalhaut	1	11. 0. 34. 47	21. 6. 28 S
* α Pegaſi	2	11. 20. 14. 30	19. 24. 38 N

N. B. Thoſe Stars only marked with Aſteriſcs are made uſe of in the Diſtances of the Aſtronomical and Nautical Ephemeris.

B

A 1767 table of the longitude and latitude of certain stars, calculated by Charles Mason and based on James Bradley's earlier observations.

was an unprecedented database of precise and accurate observations, together with barometric and thermometric records with every observation, as well as other possible sources of error.

Bradley intended that these observations would be reduced by his successors at the Royal Observatory, as soon as a viable theory of atmospheric reduction had been developed.[49] Unfortunately, Samuel Peach, the man who acted as Bradley's executor on behalf of Bradley's daughter Susannah, denied all access to the observational registers from 1762 to 1776, during which time the Board of Longitude entered into a lawsuit to gain access to these precious observations. Peach refused to surrender Bradley's observations until he was rewarded financially.

His son, the Revd Samuel Peach, was married to Bradley's daughter, and eventually surrendered Bradley's 'working papers' to Lord North, the Chancellor of Oxford University. As Prime Minister, Lord North had more pressing matters to deal with, following the American Declaration of Independence on 4 July 1776, but gave instructions for the early publication of Bradley's Greenwich observations. The Board of Longitude withdrew its lawsuit, satisfied that the observations would soon be published – but the first volume was not published for twenty-one years, and included only the observations made from 1750 until 1755 with none of them reduced. Before this, voices were raised, not least by the Board of Longitude, about the apparent lack of urgency in publishing the observations from 1750 to 1762. Thomas Hornsby, Bradley's successor as Savilian professor of astronomy, appears to have undervalued the third Astronomer Royal's stellar observations, while the Syndics of the Clarendon Press did not consider allocating the task to another person.

Friedrich Wilhelm Bessel (1784–1846), German astronomer and mathematician. He is remembered for the Bessel functions in mathematics, and for measuring the annual parallax of the star 61 Cygni.

A second unreduced volume was much more speedily produced, edited by Abraham Robertson, later Savilian professor of both geometry and astronomy (see Chapter 5), and published in 1805. But it was not until April 1861 that the registers of the third Astronomer Royal were finally sent from the Bodleian Library to the Royal Observatory, after extensive negotiations by the seventh Astronomer Royal, George Biddell Airy. After examining the registers and comparing them with Hornsby's volume, Airy commented that 'Dr Hornsby's printed book is not an exact counterpart of the original', with his greatest censure directed not at Hornsby, but at Peach.[50] When Bradley gave over the care of his observations and other effects to his daughter's family, he could not have believed how much his trust was to be betrayed. Almost nothing of Bradley's private life was released into the public sphere, and it must be assumed that his personal papers were destroyed by his executors.

But Bradley's aims were not achieved for a further fifty years. It was only after 1807, when the German astronomer Heinrich Olbers had acquired both volumes of Bradley's published observations and given them to Friedrich Bessel, that their true riches were revealed. Olbers had believed that the young Bessel possessed the skills and energy to reduce the treasure trove of rigorous observations made by Bradley and his assistants from 1750 to 1762. For twelve years Bessel worked on reducing this immense databank, transforming the procedures leading to the reduction of astronomical data, and published it as a catalogue of 3222 stars in his *Fundamenta astronomiae*, describing Bradley and his assistants as *viri incomparabilis*.[51] In this way, Bradley's astronomical work laid the foundations for 18th-century astrometry.

Bessel firmly believed that reduction should always accompany observation. This was not a convenience that was available to Bradley, who lacked confidence in applying his ad hoc theory of atmospheric refraction to the reduction process. Bradley had developed an approach based on extended observations over many years, but was unable to vouch for this approach in all conditions, preferring instead to keeping careful records of barometric and thermometric observations. In the high-precision science of astrometry, he felt unable to apply such estimates, and chose instead to leave a database that would allow his Greenwich successors to carry out the reductions in accord with the now well-developed theory of atmospheric refraction. It must also be acknowledged that, during his final years, Bradley no longer retained his earlier strength and energy, often living with abdominal pain. Even Bessel, a young and ambitious astronomer, took twelve years to reduce the observations that had been made by Bradley and his assistants.

Bradley's catalogue takes up only half of Bessel's *Fundamenta astronomiae*, and this book transformed the science of astrometry. As well as revealing the content of the observations made by Bradley and his assistants from 1750 to 1762, this vast database of high-precision observations led Bessel to resolve many problems associated with the reduction of data. Bradley had incorporated many more variables than his predecessors, each of which had to be reduced separately. Bessel also incorporated the 'personal equation' denoting the minute differences in the observational practices between Bradley and his three assistants.[52] The increased precision of their work over that of Bradley's predecessors exposed the manifold problems involved. Since Bradley's death in 1762 the problems connected with reduction had multiplied, and as the precision of observations increased, the associated problems became unmanageable. Given the spread of values given to constants such as aberration or nutation, few considered whether the derived values were actually correct. Indeed, Bessel recognized that they were all likely to be inaccurate, and while reducing the Greenwich observations he assessed their accuracy. In particular, Bessel for the first time challenged the notion that a constant, such as that for aberration, possessed a specific value.

In his *Fundamenta*, Bessel calculated errors at each succeeding stage of his work. By carrying the errors through each phase of the calculations, he attempted to make the final errors more credible than earlier estimates, with the errors accounted for as rigorously as the data would allow. The work undertaken by Bessel opened the doors to advances in the science of positional astronomy. Since Bradley's death, observations had become ever more precise, and the problems of reduction first encountered by Bradley in the 1750s had become increasingly pressing. The Greenwich observations described as *incomparabilis* by Bessel were so, because in Bradley he encountered an astronomer who was as concerned with the sources of error as he was. Bessel derived a methodology for calculating the parameters of these limits, but regrettably, the avarice of Bradley's executor and the indolence of Bradley's Oxford successor denied access to his registers. The silver lining was, of course, that instead of Bradley's observations being reduced by less astute successors at Greenwich, they had finally come into the hands of a mathematical genius, who used Bradley's work to transform the science of positional astronomy just as surely as Bradley had achieved in the 18th century. In 1838, Bessel was able to measure the annual parallax of the star 61 Cygni, for the first time allowing stellar distances to be ascertained by direct geometrical techniques, now the bedrock foundation of all astronomical distance scales.

Interior of the Radcliffe Observatory, built in the 1770s at the instigation of Thomas Hornsby. It was the first academic establishment in Europe to combine teaching and original research in astronomy.

CHAPTER 5

Enlightenment professors

ROB ILIFFE AND SARAH DRY

This chapter describes the life and work of the three men (Thomas Hornsby, Abraham Robertson, and Stephen Peter Rigaud) who held the position of Savilian Professor of Astronomy between 1763 and 1839. Astronomy remained a relatively popular subject at the University in the late 18th and early 19th centuries, and it was taught to a surprisingly large number of students. However, at the time of James Bradley's death it was difficult to fulfil that part of the Savilian statutes that enjoined the teaching of practical astronomy.

Thomas Hornsby, Bradley's immediate successor, recognized the need for a new building in which to teach and make observations, and in the late 1760s he led a campaign to build a world-class observatory. The Radcliffe Observatory briefly made Oxford the leading and best-equipped locale in the world for conducting positional astronomy, but by the end of the 1830s its major instruments needed replacement. Editing duties, particularly the curation and publication of Bradley's data, loomed large in the careers of Robertson and Rigaud, and although these obligations were obviously significant as a means of providing reliable data for the astronomical community, they seriously curtailed the ability of the Savilian professors to produce original theoretical work.

Introduction

The successive tenures of the three Savilian professors of astronomy following James Bradley – Thomas Hornsby (from 1763 to 1810), Abraham Robertson (from 1810 to 1827), and Stephen Peter Rigaud (from 1827 to 1839) – witnessed major intellectual and political changes both within and beyond the University of Oxford. The late 18th-century University was resolutely devoted to defending the status of the Anglican Church and remained focused on teaching Latin and Greek classical texts, along with traditional approaches to logic and ethics. Nevertheless, for those students who were interested in science, instruction was available in mathematics, anatomy, physiology, botany, chemistry, experimental philosophy and general natural philosophy, the last via lectures given by the Sedleian professor. From the early 1790s the University redoubled its efforts to support the British Empire and religious orthodoxy, and it steeled itself

Rob Iliffe and Sarah Dry, *Enlightenment professors.* In: *Oxford's Savilian Professors of Astronomy.* Edited by: Robin Wilson and Steven Balbus, Oxford University Press. © Oxford University Press (2025). DOI: 10.1093/oso/9780198894292.003.0005 to ensure that students were not seduced by subversive intellectual and political notions. In the first three decades of the 19th century, chairs were created in scientific subjects and reformers brought about a major overhaul of the examinations system, which was intended to produce a new type of student who could better serve the national interest along with Anglican orthodoxy.

The social and intellectual lives of the Savilian professors of astronomy do not seem to have been directly affected by the great geopolitical events that took place in their various tenures, although they taught many individuals who would become major political figures in 19th-century Britain. The various demands that the Savilian statutes placed on them – giving regular lectures on astronomy, practising positional astronomy at the new Radcliffe Observatory (from 1774), and publishing reliable and useful observational data – necessarily absorbed most of their attention. Although, they were aware of, and in many cases taught, the great scientific discoveries that were transforming the broader intellectual worlds in which they worked, the subject of astronomy remained an intellectual sanctuary, being largely impervious to contamination by emerging materialist or evolutionary ideas.

None of the figures discussed in this chapter could hope to match Bradley's theoretical achievements, but Thomas Hornsby, Bradley's immediate successor, maintained his predecessor's commitment to meticulous observation, especially when the new Radcliffe Observatory became operational. Astronomy remained a relatively popular subject at the University, and in the nearly eight decades covered by the tenures of these three individuals it was taught to a surprisingly large number of students. The construction of the Observatory, promoted and overseen by Hornsby, briefly gave Oxford the leading and best-equipped observatory in the world, but by the end of the 1830s its major instruments needed replacement. The curation and publication of Bradley's work lingered among the duties of the professors up to the 1830s, but they nevertheless continued to produce meticulous observations, coupled with the recording of thermometric, barometric, and rainfall data.[1]

Thomas Hornsby

Hornsby was born in Durham in 1733 and matriculated in Oxford at Corpus Christi College in December 1749, becoming a fellow of the college in 1760. He attended a course on experimental philosophy given by Bradley in 1751 and in the following years he must have demonstrated his prowess in the several branches of astronomy to Bradley and to Nathaniel Bliss, the Savilian professor of geometry and Bradley's successor as Astronomer Royal. Hornsby benefited considerably from their support; indeed, throughout his career he showed a pronounced ability to attract powerful patrons.

With the help of these astronomers, by 1755 Hornsby had become acquainted with George Parker, 2nd Earl of Macclesfield and president of the Royal Society. Hornsby observed the June 1761 transit of Venus at Macclesfield's observatory at Shirburn Castle, and in the three decades following the Earl's death in 1764 he made several major observations there with Macclesfield's heir and 3rd Earl, Thomas Parker. At the end of the 1760s Hornsby also acquired the support of another keen aristocratic astronomer, George Spencer, 4th Duke of Marlborough. Hornsby worked closely with Marlborough for decades at his Blenheim Palace observatory, offering him instruction in astronomical practice and providing advice on the purchase of equipment. The backing of Macclesfield and Bliss was crucial for Hornsby's assumption of the Savilian chair and for his election to a fellowship of the Royal Society, both of which took place early in 1763.

This portrait of Thomas Hornsby (1733–1810) hangs in Green Templeton College, Oxford, where the former Radcliffe Observatory is located.

His prospects were such that he was able to marry a few months later, and he moved to the Savilian professor's house in New College Lane, which for some decades had adjoined the residence of the Savilian professor of geometry (see Chapter 4).[2]

Thomas Hornsby was exceptionally active in the first decade of his post. In addition to his role as Savilian professor, he also inherited the role of Reader in Experimental Philosophy from Bradley, giving at least two lecture courses every year for nearly four decades. In 1782 he would also become the Sedleian professor of natural philosophy, and in the following year he was appointed Radcliffe Librarian. These other pursuits took up a sizeable portion of his time, but his roles as Savilian professor (and later Radcliffe Observer) consumed most of his energies over a career that lasted nearly half a century. On his appointment, Hornsby sat on the Board of Longitude and attended numerous meetings on the best methods for determining longitude at sea. Concern with this problem reached its zenith in the 1760s, and navigators were presented with the opportunity to test the relative utility of the so-called lunar method and the chronometer, the latter improved over several decades by John Harrison. From 1764 Hornsby supported the Astronomer Royal Nevil Maskelyne in his efforts to promote the lunar method as the technique of choice, although calculations using tables provided in the *Nautical Almanac* were excessively tedious and sailors preferred to use a chronometer.[3]

Hornsby tended to observe the heavens during the daytime and his early published work, written in his first decade as Savilian professor, was focused on observations of the Sun, the size of the solar system, and the proximity of near stars. His first paper, read at the Royal Society on 23 December 1763, concerned the measurement of the solar parallax, the angle subtended by the Earth's equatorial radius at the centre of the Sun at 1 Astronomical Unit. Knowing the diameter of the Earth at the equator, solar parallax could be determined by comparing the duration of the recent 1761 transit of Venus observed at different places with the notional duration of the event as seen at the centre of the Earth. Hornsby noted that the determination of solar parallax could take place only with the best observations 'made in places where longitudes are as accurately ascertained as the present state of Astronomy will permit'. When this had been done, one could critically assess the times for the ingress and egress of Venus across the Sun's disc given by observers at different sites around the globe. The bulk of Hornsby's paper consisted of a detailed comparison of various observations of the transit, and he concluded that the recalibrated data from these observations could not yet give the level of accuracy for solar parallax that astronomers required. He drew the attention of his readers to the imminent transit 'pair' of 3 June 1769, and argued that properly conducted observations should produce a value for the solar parallax to within 1% of the true figure.[4]

In February 1766 the Royal Society was treated to a comprehensive paper by Hornsby on the best location to view the forthcoming transit. Noting that a similar opportunity would not recur until 1874, Hornsby proposed a number of possible observational locations around the globe, devoting a substantial part of his exposition to an analysis of accounts by Dutch, French, and Spanish explorers of possible land masses in the Pacific Ocean. Due in no small part to the impact of Hornsby's paper, observing the transit became a central scientific goal of James Cook's first expedition. Cook and the *HMS Endeavour* astronomer Charles Green set out to use the newly discovered island of Tahiti as their base for measuring its duration, and they successfully observed the phenomenon on 3 June 1769. Hornsby himself viewed the transit in Oxford from the top of the mathematics tower in the Schools Quadrangle. Just over a week later he sent a paper to the Royal Society, detailing the observations that he and others had made at Oxford. To these he added the results of observations made at Shirburn by the 3rd Earl of Macclesfield and his assistants, notable among whom was Lady Macclesfield.[5]

Hornsby initially used a 12-foot refractor to see the moment that the planet first touched the outer edge of the Sun, switching to a 7½-foot refractor with an achromatic lens made by Peter Dollond (with 90× magnification) as it started to cross the Sun's disc. Hornsby mentioned that the Russian scholar Vasily Nikitin had observed the transit from a room at the top of the Radcliffe Infirmary, and he wrote a testimony regarding Nikitin's skills for him to take back to St Petersburg. On the following day (4 June) Hornsby observed a solar eclipse from the same location, noting that the eclipse and the transit had also been observed at Oxford by Samuel Horsley and Cyril Jackson, both men (according to Hornsby) 'not less distinguished by their zeal for astronomical and mathematical inquiry, than for their extensive knowledge and erudition'. Horsley was a fellow of the Royal Society and later the editor of Isaac Newton's collected works, while Jackson would become Dean of Christ Church and a dominant academic figure at Oxford. In the decade after 1789 Jackson attended several Lunar Society meetings in Birmingham, and was a prime mover in reforming the examinations system to place a greater emphasis on mathematics and the natural sciences. It was presumably his friendship with Matthew Boulton that led to Boulton's son attending Hornsby's Sedleian lectures in 1791, even though he was not an undergraduate.[6]

In December 1771 the Royal Society heard a major paper from Hornsby on solar parallax, in which he derived an improved figure as the result of analysing many measurements taken of the 1769 transit. Hornsby averred that the recent observations taken from around the globe provided an opportunity to 'obtain as accurate a determination of the Sun's distance, as perhaps

OXFORD, Nov. 1. 1766.

A Courſe of Lectures in EXPERIMENTAL PHILOSOPHY will begin to be read at the *Muſeum* on *Monday* the 10*th* Day of *November* at half an Hour after Two o' Clock in the Afternoon.

OXFORD, May 9. 1769.

THE *Savilian* Profeſſor of Aſtronomy gives this publick Notice, that he propoſes to begin a *Courſe* of *Lectures* on the TRANSIT of VENUS at the *Muſeum* on *Wedneſday* the 24th of this Month, at Three o'Clock in the Afternoon: In which the Hiſtory of former Tranſits will be delivered; the Method of computing the Places of the Sun and Planets explained and exemplified in the Caſe of the enſuing Tranſit; the Method of computing the Effect of Parallax, and of finding the Places upon the Earth's Surface, where Obſervations may be made with the greateſt Advantage, will be pointed out; and the Manner of determining the Quantity of the Sun's Parallax from ſome of the principal Obſervations made in the Year 1761 will be ſhewn and illuſtrated by Examples.

As it is propoſed to make the principal Calculations at the Time of Lecture, thoſe Gentlemen, who are deſirous of attending the above Courſe, muſt previouſly furniſh themſelves with the following Books:

Sherwin's Logarithms, 8vo.
Or, *Gardener's* ———, 4to.
Halley's Aſtronomical Tables, 4to.
Or, Tables Aſtronomiques de M. *Hallei* publieés par M. *de la Lande*, 8vo. *Paris*, 1759.
The Abbe *de la Caille's* Tables of the Sun, as publiſhed by M. *de la Lande* in his *Aſtronomie*, two large Vols in 4to. or in his *Expoſition du Calcul Aſtronomique*, 12mo. *Paris*, 1762.

Terms of Admittance One Guinea.

[*Left*] A notice of Hornsby's lectures on experimental philosophy. [*Right*] Hornby's lectures on the 1769 transit of Venus demanded a considerable level of input from the students.

the nature of the subject will admit'. He tabulated information from places as far afield as Tahiti, California, Hudson's Bay, the Kola peninsula, and Vardø in Norway, carefully and judiciously comparing the various recorded times of ingress and egress at different locations. He concluded that the mean parallax was 8.78 arc-seconds and that therefore the mean distance of the Earth from the Sun was approximately 93,726,000 miles. These figures, remarkably close to their modern values, also gave a new estimate for the expanse of the solar system, although William Herschel's discovery of Uranus in March 1781 would soon double its known size. Hornsby, who had an excellent relationship with Herschel, spent a substantial part of the year following the discovery of the planet by using an equatorial sector to track its path through the heavens, and he mentioned this in his lectures.[7]

In 1773 Hornsby published his last paper, on the quantity and direction of the proper motion of the red giant Arcturus. Edmond Halley, Jacques Cassini, and Pierre le Monnier had all argued that this star exhibited the largest proper motion (the change in observed position relative to other stars) of any known star. To ascertain the magnitude and direction of this motion more precisely, Hornsby compared more recent observations of the star with recalibrated measurements that had been produced in 1690–92 by the Astronomer Royal John Flamsteed. His own observations were performed in 1767–68 with a 43-inch transit telescope and a 32-inch mural quadrant, both made by the great London instrument-maker John Bird.[8]

Hornsby concluded that every year Arcturus moved 1.205 arc-seconds westwards in Right Ascension and 2.005 arc-seconds in Declination. He confirmed that Arcturus exhibited the largest visible motion of any star and added that it was therefore probably the closest star to the solar system visible in the northern hemisphere. For this reason it was also the most suitable

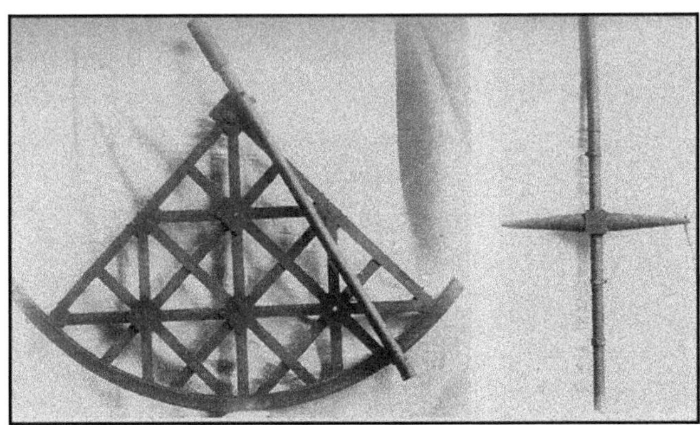

Hornsby's 32-inch mural quadrant and 43-inch transit, both constructed by John Bird. With these instruments Hornsby observed the 1769 transit of Venus and found the latitude of Oxford to be 51° 45′ 15″.

star for determining stellar parallax, the apparent change (if it could be detected) in relative positions of nearby stars against more distant stars measured at six-month intervals. Discovering the magnitude of parallax would in turn provide evidence for the real distances of stars from the solar system and ultimately shed light on the structure of the Universe, 'agreeable to the idea we may entertain of an all-powerful benevolent Creator'.

In the second half of his paper Hornsby argued that the mean obliquity of the ecliptic (the inclination of the Earth's equator to the plane of the Earth's orbit) was decreasing – a conclusion that supported Bradley's account of nutation and thus confirmed Newton's theory of universal gravitation. Hornsby's account countered claims made by Jacques Cassini and Pierre le Monnier that the obliquity of the ecliptic was unchanged, or had changed only to a negligible degree. He initially compared his own 1771 and 1772 observations with measurements taken at Shirburn, using a mural quadrant strengthened and redivided in 1745 by Bird. These, he stated, proved 'beyond all doubt, that the obliquity has become less'. But because there had been only a short interval of time between the two sets of observations, he next compared his own recent measurements with corrected earlier observations made by Flamsteed, reduced to their mean position at the summer solstice of 1690. Hornsby concluded that the decline of the obliquity of the ecliptic had amounted to 47 arc-seconds since 1690, or a rate of 58 arc-seconds per century, a result (he noted) that 'will be found nearly at the mean of the computations' produced by Leonhard Euler and Jérôme de Lalande 'upon the principles of attraction'. This remark occupied only the last three lines of his paper, but its pro-Newtonian import was clear.[9]

Founding the Radcliffe Observatory

Not long after assuming his post as Savilian professor, Hornsby had seen an opportunity to improve his own working conditions and to turn Oxford into a major centre for the teaching and practice of astronomy. In a petition presented to the trustees of the Radcliffe Trust in November 1768, he complained about the inadequate space and ageing equipment available to him, which made it impossible for him to deliver a course on practical astronomy. For these reasons (he wrote) he was seeking funding for a new and properly furnished observatory, modelled on the Royal Greenwich Observatory, that should be supplied with instruments 'as will not be equalled in the entire World'. The Observer would be the Savilian professor, who would

An engraving of John Radcliffe and a statue of him by Martin Jennings. The £40,000 trust that he bequeathed in 1714 would eventually provide Oxford University with a library (the Radcliffe Camera), an infirmary, and the Radcliffe Observatory.

make daily observations with the equipment; however, he would need an assistant, with further support for the yearly publication of the observations. Hornsby added that he was committed to delivering a course on practical astronomy that would be given twice a year, and he appealed for additional funding to supply instruments for classroom use and a 'room for experimental philosophy'. The Observer would require on-site accommodation in the form of a house, above which would be two rooms where he could pursue experimental philosophy and observe the skies across 360° by using an equatorial sector. He estimated that, taking into account the cost of purchasing land for the purpose, the whole enterprise should amount to no more than £7000.[10]

In making this petition Hornsby had already canvassed support from major grandees, including the Duke of Marlborough and the Earl of Lichfield, the latter of whom held the two posts of Chancellor of the University and Chair of the Radcliffe Board of Trustees. The trustees told Hornsby that they were well disposed to his request, but that he would have to wait until the expenses related to the construction of the Radcliffe Infirmary were settled.[11] In November 1770 the Duke leased 40 acres of land next to the Infirmary from St John's College with the intention of allowing some of that land to be used for the Observatory, and in the following year the trustees agreed to disburse sufficient funds to underwrite Hornsby's proposal in full. In 1772, having acquired the rights to use the land for astronomical purposes from the Duke, they endorsed a new plan for the Observatory in which the central building or tower was to be separated from the house. The cost of the tower was now itself estimated at close to £7000, while the instruments were assessed at £2500.

The instruments were expensive because Hornsby wanted them to be the best that money could buy, and for this purpose he drew on his close relationship with John Bird (see Chapter 4). Bird had come to the fore as a master craftsman in the 1740s, and for the next three decades he was the premier maker of brass hand-graduated instruments in Europe. He supplied Bradley

John Dollond (1706–61), maker of optical and astronomical instruments.

with several telescopes for the Greenwich Observatory in the 1740s, and made state-of-the-art brass quadrant and transit instruments for the Observatory in 1750. Hornsby was in regular communication with Bird and owned several Bird instruments, and by the time that he petitioned the Radcliffe trustees he knew that Bird was ailing and increasingly housebound. At the start of 1771 he exerted pressure on the trustees to act quickly in acceding to his request. Having received a positive response, Hornsby visited Bird at his London residence and presented him with a wish-list for the new Observatory.[12]

The most significant instruments that Hornsby requested from Bird were two 8-foot brass quadrants, each of which would have a 3-inch object glass to be made by the optical instrument-maker John Dollond, a transit telescope with a Dollond achromatic lens of 4-inch aperture, a 12-foot zenith sector, and an equatorial sector with a 5-foot focus. Bird, who had previously been sceptical about the benefits of achromatic lenses, duly approached Dollond for glasses according to these specifications, since the focal length of his glasses would dictate the radii of the quadrants. Dollond told Hornsby in October 1771 that the quadrants should magnify 60× and the transit 80×, adding that he was as keen as Hornsby was to produce achromatic lenses that would be better than anything that he had previously made.[13]

Bird initially quoted £1260 for the equipment, warning that the size of the quadrants would require a special 'wooden house' in which to divide the instrument – that is, to etch the divisions onto the scale. Hornsby's estimate of £2500 for the total expenditure on all the instruments needed for the Observatory included a number of reflectors and refractors (some of which were for teaching purposes), a smaller transit telescope and equatorial sector, and several barometers

[*Left*] By the end of the 18th century the Radcliffe Observatory was the best-equipped astronomical observatory in the world. [*Right*] The figure of Atlas supporting the globe.

and thermometers. He also requested four pendulum clocks, a chronometer constructed 'on Mr Harrisons principle to go without Oil', and a watch that could display seconds. It is unclear how many of these were to come from Bird, but as a sign of the deep friendship that existed between them Bird reassured Hornsby in February 1771 that he would ignore all other commissions to produce the exceptionally fine instruments that Hornsby required. Indeed, he revealed that Nevil Maskelyne, the Astronomer Royal, was putting serious pressure on him to produce instruments of a slightly smaller specification for Greenwich, but told Hornsby that he would 'not only, for your sake, but also for my own, refuse every thing that will retard the progress of your Instruments'.[14]

Bird worked quickly, overcoming the problem of acquiring sufficiently robust brass bars from which the quadrants would be made, and all the instruments arrived at the Observatory by the summer of 1773. Hornsby was duly impressed by the exquisite craftsmanship involved in their construction, but now had months of painstaking work ahead of him to set them up. He moved into the Observer's house in November 1773 and immediately started to put the instruments into working order. This process was complicated by the slow construction of the main tower, whose increasing weight repeatedly caused the 'stones' (piers) supporting the exceptionally heavy transit telescope to sink into the ground.

The trustees had originally chosen Henry Keene as the architect for the commission, and construction work began early in 1772. However, they were not satisfied with the external design of the main tower, and in 1773 they acquired the services of the fashionable James Wyatt.[15] The basic structure of the tower was completed by the end of 1778, although the lecture chamber (the room that Hornsby had originally requested for experimental philosophy) would not be finished for another decade. The stunning central building, whose external features were modelled by Wyatt on the Tower of the Winds in Athens, was not completed until 1798. By this time Hornsby may have been declining as an observer, but was still closely concerned with various aspects of the design – notably, the figures of Atlas and Hercules who to this day support the copper globe that marks the pinnacle of the tower.

In October 1777 the Danish astronomer Thomas Bugge visited Oxford and made a beeline for the Observatory after a short visit to the Radcliffe Camera ('Dr. Radcliffe's Library').

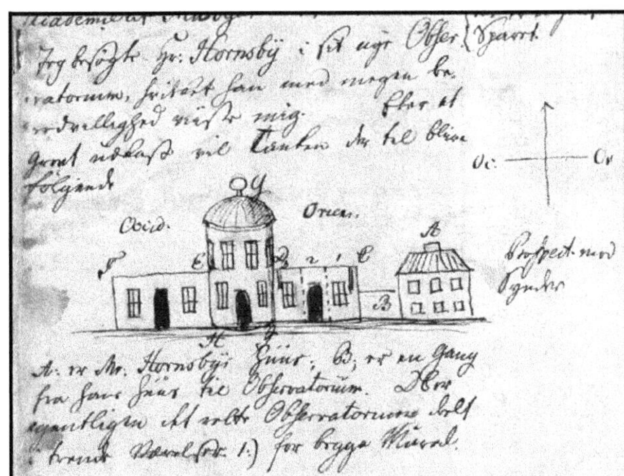

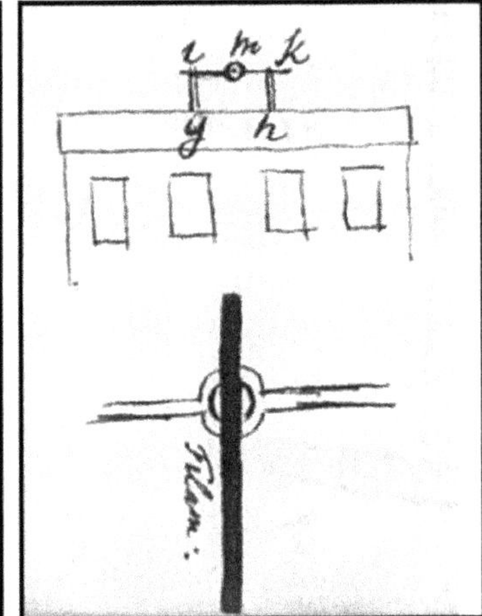

[*Left*] Thomas Bugge's sketch of the Observatory. [*Right*] Bugge's sketch of the moveable ring-shaped meridian marker on top of the Horse and Jockey pub.

Although the main tower was unfinished, Bugge drew a sketch of the Observatory as it then was, including Hornsby's house (*A*), the working or 'real' observatory (*DC*) (now the 'east wing') reached via corridor (*B*), the central tower (*GHI*), and the wing (*FE*) that was to contain the smaller set of instruments which Hornsby had used until recently. The latter was to be for the use of students in connection with his course on practical astronomy. The second floor of the working observatory was further divided up into three parts: room 1 which housed the mural quadrants, room 3 which contained the transit, and an intervening room which held the zenith sector.

Bugge gave only a short account of the quadrants, comparing them with the descriptions of the same instruments given in Bird's 1768 treatise on their construction, which Bugge had bought in London earlier in the year. He recorded that there were several smaller Dollond telescopes in the same room as the quadrants, including one that belonged to Hornsby himself, and noted the presence of several barometers and rainwater gauges, as well as four thermometers graduated according to the Fahrenheit scale, two of which were made by Bird.[16]

In the central room stood the 12-ft zenith sector, mounted on a stone block that sat in a solid gravel base. Bugge recorded that the instrument was 'constructed according to special principles so that, without doubt, it is the best of all sectors'. The transit instrument in room 3 had an 8-ft achromatic refractor and Bugge was particularly impressed by it, describing its mechanism in great detail. He noted that because it was counterbalanced (with both weights marked 'p'), he could lift the instrument out of its bearings with one finger. Reversing the instrument, which weighed over 100lbs, required two men to lift the axis entirely out of its bearings and turn it through 180°.

Thomas Hornsby told Bugge that the apparatus was so sensitive that if he touched it, the heat of his hand would cause the axis to expand, moving it out of alignment with the meridian marks that he had erected on two buildings to the north and south of the observatory. One meridian marker was placed on a wall at Worcester College 680 yards to the south, while to the north Bugge viewed a moveable ring-shaped marker on the roof of the Horse and Jockey Pub on Woodstock Road, noting that the 'central hair of the instrument divided the hole exactly'.

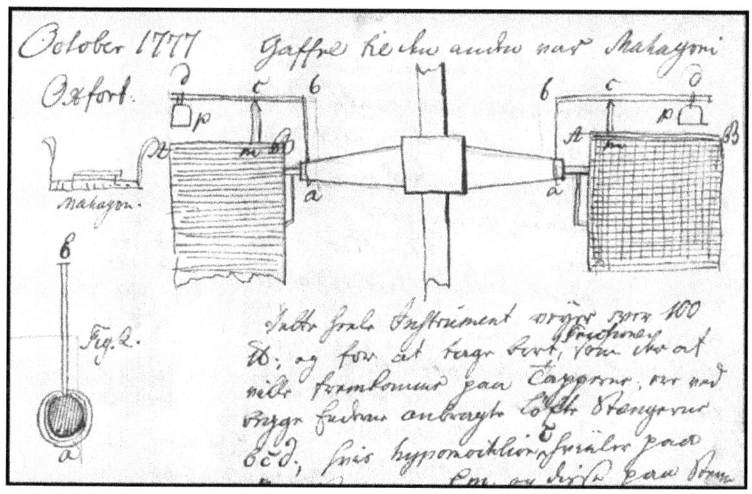

Bugge's sketch of the transit in room 3.

Hornsby allowed Bugge to use the instrument, both during the day and at night, and Bugge was convinced that it could produce positional observations with an accuracy of about 0.2 arc-seconds. He added that its magnification was so great that he had initially considered such a power 'as boastfulness rather than reality', but experience had taught him otherwise. After a week in Oxford Bugge left the Observatory with regret, remarking in his journal that 'both as regards the arrangement and the instruments', it was without doubt the best in Europe.[17]

Hornsby as lecturer and observer

The statutes of the Savilian chair placed lecturing obligations on the incumbent. Bradley had delivered his lectures on experimental philosophy and astronomy in the Old Ashmolean Museum, but in astronomy he had not attempted 'hands-on' lectures where students could use equipment. As we have seen, Hornsby's interest in offering lectures in practical astronomy had been a central motivation for his 1768 petition to the Radcliffe Trustees, but the glacial pace of the construction of the Observatory's lecture room thwarted his efforts to teach the art of astronomical observation alongside a more theoretically inclined approach.

Hornsby left an extensive record of his Savilian lectures in the form of presentation notes; these indicate that their content remained fundamentally the same over nearly four decades, except for some occasional updates. Bugge recorded in October 1777 that Hornsby gave a course of fourteen lectures, each of which lasted one hour and three-quarters, and he added that Hornsby had later presented him with a printed outline of their content.[18] The outline describes the standard format in which Hornsby taught, although occasionally he gave an ad hoc course of lectures on unusual celestial events, such as the 1769 transit of Venus.

Hornsby began his series of lectures by tracing the history of astronomy, occasionally discussing where celestial topics had been mentioned in classical writings. With the help of an assistant he made copious use of 'schemes' (maps and diagrams) and astronomical models, indicating in his lecture notes the places at which a celestial map or the frequently deployed

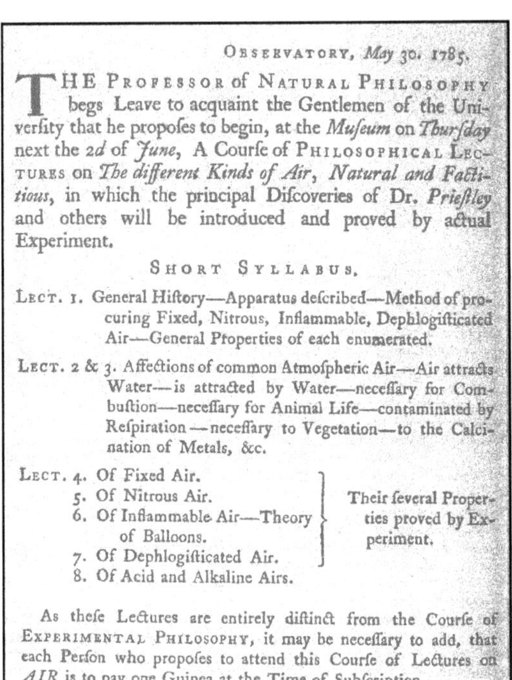

OBSERVATORY, *May* 30. 1785.

THE PROFESSOR of NATURAL PHILOSOPHY begs Leave to acquaint the Gentlemen of the Univerſity that he propoſes to begin, at the *Muſeum* on *Thurſday* next the *2d* of *June*, A Courſe of PHILOSOPHICAL LECTURES on *The different Kinds of Air, Natural and Factitious*, in which the principal Diſcoveries of Dr. *Prieſtley* and others will be introduced and proved by actual Experiment.

SHORT SYLLABUS.

LECT. 1. General Hiſtory—Apparatus deſcribed—Method of procuring Fixed, Nitrous, Inflammable, Dephlogiſticated Air—General Properties of each enumerated.

LECT. 2 & 3. Affections of common Atmoſpheric Air—Air attracts Water—is attracted by Water—neceſſary for Combuſtion—neceſſary for Animal Life—contaminated by Reſpiration—neceſſary to Vegetation—to the Calcination of Metals, &c.

LECT. 4. Of Fixed Air.
5. Of Nitrous Air.
6. Of Inflammable Air—Theory of Balloons.
7. Of Dephlogiſticated Air.
(4–7: Their ſeveral Properties proved by Experiment.)
8. Of Acid and Alkaline Airs.

As theſe Lectures are entirely diſtinct from the Courſe of EXPERIMENTAL PHILOSOPHY, it may be neceſſary to add, that each Perſon who propoſes to attend this Courſe of Lectures on *AIR* is to pay one Guinea at the Time of Subſcription.

An advertisement for Hornsby's lectures as Sedleian professor of natural philosophy.

orrery might be used to illustrate a point. He discussed an impressively wide range of current astronomical theories (such as the nature of the Sun's atmosphere), but as a good Newtonian he was careful to distinguish hypothetical claims from conclusions reached by calculations based on reliable observations.

As a conclusion to his courses on astronomy, Hornsby offered his students the opportunity to learn the 'principles of astronomical calculation' by using a lunar eclipse as a case study. Those attending were supposed to bring along Mayer's lunar tables and a book of logarithms to the sessions, a fact that demonstrates the substantial level of technical skill Hornsby expected his students to possess. Although he could not yet give his students practical instruction in the use of instruments, Hornsby introduced them to the most advanced issues in astronomy, while instructing them on more standard topics such as fixing the date of Easter and assessing the relative advantages of the Julian and the recently introduced Gregorian calendars. Newtonian physics dominated the study of astronomy at this time, and he discussed the central accounts in the *Principia mathematica* relating to the nature and causes of tides, lunar motion, the oblate spheroidal Earth, and the paths of comets. He also discussed Bradley's key discoveries of aberration and nutation and, as noted above, he periodically added references to major discoveries by Herschel and others.[19]

According to Henry Best, who attended Hornsby's Sedleian lectures in the late 1780s, his 'mode of instruction was peculiarly clear, his language correct, with choice phrases and well-turned periods', although Best felt that the content of his lectures was overly pedantic.[20] He added that Hornsby suffered from epilepsy, and that having been helped to a chair by a servant while suffering from a fit, he would then resume where he had left off once the seizure

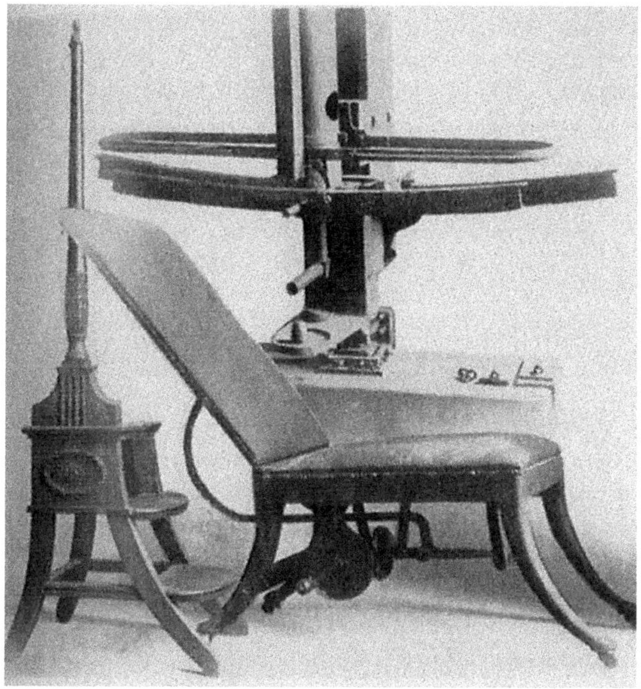

John Bird's 12-foot zenith sector at the Radcliffe Observatory, with its observing chair.

had ended. This fact may at least partly explain why Hornsby found it so difficult to fulfil the editorial duties related to his role in the publication of Bradley's data (as discussed below).

In the 1750s Hornsby had honed his observational skills at Corpus Christi College and at the tower of the Schools, and from 1763 he was able to make use of the small observation room at the top of his New College Lane house. He used his own observations in the later papers that were published in the *Philosophical Transactions*, but it is ironic, if understandable, that his career as an original scientific author ceased when he moved to the new Observatory. Once the quadrants, transit, sector, and clocks had been installed there, Hornsby was in charge of the best-equipped observatory in the world, and was also in an ideal position to make observations that could surpass those of Bradley for precision and accuracy. Moreover, in principle, each instrument could be used as a check on at least one of the others; for example, quadrants were used in conjunction with the transit telescope, while the zenith sector checked the vertical alignment of the quadrants against the zenith stars in the Draco constellation.[21]

The continual checking and recalibration of these instruments, coupled with 'keeping the clock' – that is, taking accurate timings of observations – would ordinarily have required the labours of at least one assistant. However, Hornsby lacked any personnel support throughout his almost fifty years as Savilian professor, a particularly remarkable fact given the health problems to which he was increasingly prone in the latter stage of his career.

Hornsby's records provide extensive insight into how he actually produced his observations. The transit instrument and the quadrants had five vertical wires (with two on either side of the meridian wire) and one horizontal wire. In order to catch the declination of the star in the quadrant instrument as it passed the meridian, he would have to leave the transit telescope after the star had passed the first two wires to the right of the middle wire; he would then hurry back to the transit instrument to view the star passing through the last two wires.

All this, as Allan Chapman has noted, required tremendous dedication, an excellent memory, and unusual physical stamina.[22] Although, as we shall see in the next section, Hornsby devoted much of his working life as Radcliffe Observer to the publication of Bradley's data, subsequent analysis of his own work (notably, his transit observations) shows that in many ways, aided by better-quality instruments, it surpassed the accuracy of his illustrious predecessor.

Editing Bradley's papers

If single-handedly carrying out the many activities demanded of him as Radcliffe Observer were not enough. In the mid-1770s Hornsby was given the responsibility for overseeing the publication of observations made at Greenwich by Bradley in his position as Astronomer Royal. The fate of these papers in the aftermath of Bradley's death constitutes a complex familial and institutional saga. The Royal Society, and then the Board of Longitude, had a serious interest in procuring the observations to publish lunar tables that would be of use to astronomers and to navigators who needed them for the determination of longitude by the lunar method. However, Bradley's executors had taken the papers from the Greenwich Observatory soon after his death and (as we saw in Chapter 4) they came via his daughter Susannah into the possession of her maternal uncle Samuel Peach.[23]

From 1767 the Crown took legal action to retrieve the papers, and this action lasted until 1776 when the family gave permission to Oxford University to publish them. Hornsby evidently had the trust of the Peach family, and this must have played a role in the fact that the documents ended up in Oxford, on condition that the observations be published by the Clarendon Press. It fell to Hornsby to ensure that they were made public, although it would be over two decades before they finally appeared in print.

Under continued pressure from the Board of Longitude, the Royal Society, and Nevil Maskelyne (as Astronomer Royal), Hornsby laboured to bring to the press what would be the first of two published volumes of Bradley's observations. Indeed, he was present at many of the meetings of the Board of Longitude where the delay in the publication was condemned as unacceptable. The first volume finally appeared in 1798, and in his introduction to it Hornsby defended both the public-spirited intentions of the Peach family and his own role in the long delay in its appearance. Many (he wrote) had been witness to his 'assiduity and unremitted diligence' in prosecuting this 'arduous and important undertaking', and it would have been completed more quickly if it had not been for his illnesses, which may themselves have been brought on by the intensity of his devotion to the cause. His 'generous Employers' had trusted him to complete the task, and he was not going to question their wisdom in doing so.[24]

In his introduction, Hornsby also apologized for not expanding and correcting the catalogue of 389 fixed stars from 1st to 7th magnitude, and noted that he had published only a portion of Bradley's observations of meridian transits (up to December 1755) and southward meridional distances of the planets, Sun, and stars from the zenith (up to December 1758). As noted below, the Savilian professor of geometry, Abraham Robertson, would publish most of the later series of observations (up to 1762) in the second volume of the *Observations*, which appeared in 1805.

According to Henry Best, Hornsby was highly respected as 'an able and scientific man' who commanded affection from friends and respect from strangers. Towards the end of his career he was no longer able to take observations or give lectures. George Cox, the Oxford's bedel in medicine and arts for over half a century, witnessed one of Hornsby's last public acts, which was to prevent a collection of statues being stored in the Radcliffe Camera. When this move was debated in Convocation, Hornsby, who according to Best was 'rather a monopolist of academical appointments' stood up in his position as Radcliffe Librarian and brandished the key to the library, saying that with this key he now vetoed the deposit of the materials in the Library. Daring as this act of resistance was, it was a far cry from the ambitious actions that he had undertaken many decades earlier in setting up the Radcliffe Observatory.[25]

Abraham Robertson

Installed in 1810, the new Savilian professor of astronomy, Abraham (Abram) Robertson, was well known to Hornsby. Born into a humble family in Dunse, Berwickshire, Robertson attended school in Great Ryle in Northumberland before going to London in the mid-1770s, where he sought work with the East India Company.[26] Failing to find any, he ended up in Oxford, where he made an unsuccessful attempt to establish an evening school for mechanics before working as a domestic servant for John Ireland, a local apothecary. Ireland later recalled one evening when he had been discussing a problem in Euclid with a guest while Robertson stood by the sideboard, ready to serve. Calling for wine ('the topic being somewhat dry'), Ireland noticed that Robertson 'seemed deaf and distracted'. The next morning, Robertson apologized, explaining the source of his inattention, and according to Ireland he explained his behaviour thus:

> Well, Sir, if he wad ken the truth, I was deep in that problem, but had nae courage enoof to tell ye, that he were baith quite wrong.

This exchange led Ireland to persuade Christ Church to offer to Robertson a servitorship, a position where a student received free accommodation and was exempted from paying fees for lectures in exchange for acting as a servant to a college fellow.[27]

Robertson was taken under the wing of John Smith, the Savilian professor of geometry, and matriculated at Christ Church on 7 December 1775, graduating with a BA degree in 1779 and an MA degree in 1782, the same year that he became a college chaplain. Having taken up his post in 1766, Smith was frequently away from Oxford working as a physician at Cheltenham, and from 1784 Robertson delivered Smith's mathematical lectures.[28]

His obituarist noted that Robertson went out of his way to help students to come to grips with the material, loaning them his papers on many occasions. Seeing his beginners struggling to understand a particularly knotty passage in Euclid (the fifth definition in Book V), Robertson was inspired to write a complete demonstration of it for them in 1789, and a printed version of the work appeared in 1804. In 1792 he published an exposition on conic sections, a work that was dedicated to Cyril Jackson and which contained a substantial history of the general subject. In the same year he oversaw the publication of Giuseppe Torelli's edition of Archimedes, and with strong support from Oxford patrons, his work on this edition gained him membership of the Royal Society in 1795, and the Savilian chair of geometry two years later.[29]

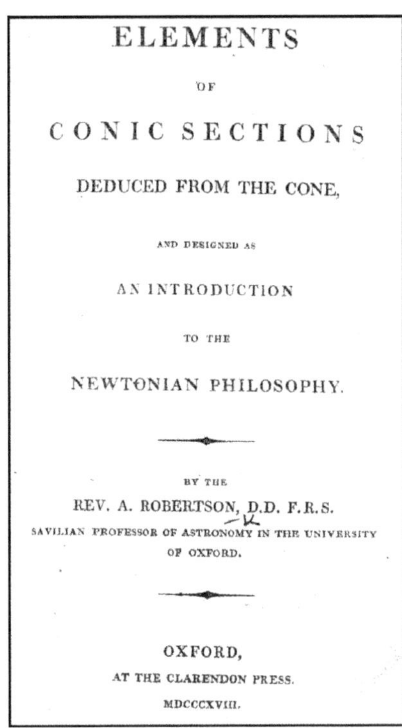

ELEMENTS

OF

CONIC SECTIONS

DEDUCED FROM THE CONE,

AND DESIGNED AS

AN INTRODUCTION

TO THE

NEWTONIAN PHILOSOPHY.

BY THE

REV. A. ROBERTSON, D.D. F.R.S.

SAVILIAN PROFESSOR OF ASTRONOMY IN THE UNIVERSITY

OF OXFORD.

OXFORD,

AT THE CLARENDON PRESS.

MDCCCXVIII.

[*Left*] Robertson's *Elements of Conic Sections* was the most influential geometry textbook at Oxford in the early 19th century. [*Right*] One of the most successful students of the time was Robert Peel (later prime minister) whose tutor used to recall that, when Peel was examined on Robertson's *Conic Sections*, the way in which he answered called forth the admiration of all that heard him.

Robertson as editor

Robertson had already been faced with a more intractable editorial problem. Following the chance discovery in 1784 of a cache of Thomas Harriot's manuscripts in Petworth House, where they had been inadvertently stored among the stable accounts, Hornsby had suggested that Oxford University Press should publish an edition of the most important papers, along with a biographical memoir of Harriot. Due to Hornsby's other commitments, Robertson was asked to evaluate the batch of disordered manuscripts, and for over a decade the Republic of Letters awaited the imminent appearance of a major edition of Harriot's works. In the end, to the chagrin of many, Robertson concluded that Harriot's papers were unsuitable for the press, and they were set aside. The project was later reactivated under the care of Stephen Peter Rigaud, Robertson's successor as Savilian professor of geometry, and in due course his successor as the professor of astronomy.[30]

When John Smith died in 1797, Robertson was in prime position to assume his place as Savilian professor of geometry and he was duly appointed to the post, holding it until 1810. Although, as we have seen, he was an attentive lecturer in mathematics, much of his time was taken up in producing the second volume of the edition of Bradley's observations. Robertson applied the skills he had acquired in scrutinizing Harriot's papers to good effect, extending the data begun by Hornsby to the end of Bradley's observing career. He performed many thousands of calculations, transforming into degrees, minutes, and seconds the figures resulting from Bird's practice of dividing the quadrant into 96 parts.

When Robertson's volume finally appeared in 1805, thousands of Bradley's unreduced observations were newly available for astronomers to examine, enabling them to consider the possible distorting effects of aberration, nutation, and a series of meteorological variables. As we saw in Chapter 4, the data would form the basis of the work of Friedrich Bessel who, with the support of the Board of Longitude, spent a decade in reducing Bradley's observations. In Bessel's great star catalogue of 1818, the *Fundamenta astronomiae*, he derived the positions of over 3000 stars at the epoch of 1 January 1755 to within $^1/_{10}$ of an arc-second.[31] With such precision, 19th-century astronomers could begin to deduce the real distances of hundreds of stars from the solar system, thus providing a much better understanding of the structure of the Milky Way.

The scale of the work undertaken by Robertson on the second volume of Bradley's observations was considerable. In his preface, he pointed to the diligence of the observers who had generated the records and the scale of the task involved in bringing them to publication. The combination of observation, registration, and publication which were required to produce robust and useful astronomical knowledge was a matter of both national and professional pride for Robertson. Links between the Clarendon Press, the Radcliffe Observatory, and the Board of Longitude enabled the publication of astronomical records whose value Robertson understood to be both theoretical and practical. As he put it, the publication would extend the mathematical material necessary for 'improving the connection between practical and physical astronomy, so happily established by Isaac Newton', a connection which both elevated the status of astronomical science and contributed, by means of anticipated improvements to navigation, to the saving of lives and property at sea. Robertson noted that the publication was the result of many long hours of work by Bradley, and that his work deserved the 'thanks of every man of science' and the 'gratitude of every maritime Nation'.[32]

Robertson as observer and lecturer

On Hornsby's death, Robertson was elected as his successor to the Savilian chair in astronomy, a move that was soon followed by his appointment as Radcliffe Observer. In his first meeting with the trustees of the Observatory, Robertson reported on the state of the instruments, largely neglected for the period of Hornsby's final years, and requested funds for some repairs. Finally, an assistant observer was hired and given lodging in the Observer's house. Unlike Hornsby, who had received a comfortable income from his multiple appointments and did not request a salary from the Trustees, Robertson lacked additional sources of income. The Trustees duly increased his salary to £300 from midsummer 1811, with the proviso that he should live at the Observatory, make and faithfully record observations, and have them printed annually.[33]

Two years later, a 10-ft Herschel reflector, equipped with a speculum finished by Herschel himself, was purchased by the trustees for 300 guineas and set up in the upper room of the Observatory building. Robertson also continued to make the meteorological observations that Hornsby had begun, allowing later astronomers to correct observations for refraction. Keeping the meridian line unobstructed also required his attention. When a new chimney was built on the Horse and Jockey pub, the landlord accepted five guineas from the Trustees to rebuild it so that it would not block the meridian line.

The testimony of a student, George Chinnery, who studied for the BA degree at Christ Church between 1808 and 1811, provides excellent evidence regarding Robertson's activities as a lecturer. Chinnery, who wrote to his mother on an almost daily basis over a period of four years, found that the study of Newton demanded extreme effort and long hours. Before Robertson began his 1810 Hilary Term mathematical lectures on Book I of Newton's *Principia*, Chinnery met him at his house on New College Lane. Robertson read out the course prospectus and advised Chinnery to study the first part of the *Principia* by himself. Chinnery, a talented classicist, told his mother that a key difficulty was that the book was written 'in a style quite particular to the subject, concise & replete with technical words hardly known by a Horace, a Virgil or a Juvenal'. Following Robertson's first 'Newtonic lecture', Chinnery told her that Newton 'is a greater devourer [of time], and a quest not easily got rid of'. A week later, on 22 February, he reported again on the rigours of the subject, expressing his gratitude for Robertson's slow delivery, which enabled him to stay focused for the full 90-minute lecture:[34]

> Robertson's 4th lecture took place today; there is one difficulty attending them, from their being public, that if in a demonstration (which he does upon a slate before us) there should be any point not immediately clear, one cannot stop him & beg that he will begin over again; this however signifies but little; it obliges to unremitting attention during the hour & a half which is the time allotted to the lecture, and complete really undivided attention will prevent any part of his demonstration being unintelligible; he is a delightful beautifully slow, beautifully clear lecturer . . . I like him much.

In 1813 Robertson began his course of astronomical lectures with a description of the Ptolemaic and Tychonic systems. Explaining that it was a 'waste of time' to follow their reasoning too closely, he moved quickly to a description of the heliocentric system, which had been the result of Copernicus's search for a theory 'more reasonable and consistent with the simplicity of nature'. He emphasized to his students the importance of observations as a check on, and confirmation of, what would otherwise be mere theorizing. While Copernicus's 'whole system was harmonious and agreeable to observation, Robertson noted that Copernicus himself was initially 'not satisfied with general correspondence between theory and observation' and had made 'observations on the planetary motions for nearly thirty-six years before he published his new system'. The truth of the Copernican theory had only finally been established by observations 'made in most Kingdoms in Europe with the greatest skill and care'.[35]

Robertson displayed the same commitment to precision and accuracy as Bradley and Hornsby had done, noting the different ways in which external conditions could affect an observation. He instructed his 1813 audience that

> the first object which an Astronomer has in view is to ascertain the real situation of a heavenly body with reference to the meridian & horizon or zenith.

Noting the distorting properties of the atmosphere, he gave instructions on how to calculate the refraction of the atmosphere for any given latitude. He also explained that the variation in the atmosphere caused by temperature and pressure changes accounted for the twinkling of the stars, making them appear 'agitated with a kind of trembling', and said that this was why 'the heights of the mercury, in the two instruments [in the Observatory] have, for more than half a century, been registered with observations requiring a correction for refraction.'[36] Robertson and his assistant took observations throughout his tenure, faithfully carrying out the measurements and checks he recommended. However, as in the case of his successor Stephen Peter Rigaud, he lacked the time to reduce his own observations, and they remain unpublished.

Stephen Peter Rigaud

Robertson died in December 1826 and Rigaud, who had succeeded him as Savilian professor of geometry in 1810, was elected in his place to the vacant chair of astronomy and the position of Radcliffe Observer. Rigaud's mother was the daughter of the eminent Huguenot Newtonian natural philosopher Stephen Demainbray, King's Astronomer at the Kew Observatory from 1768 to 1782, while his father was James Stephen Rigaud, Demainbray's assistant and his successor as Observer at Kew.[37]

Stephen Peter Rigaud matriculated as a 16-year-old at Exeter College in 1791, receiving BA and MA degrees in 1797 and 1799. Perhaps unsurprisingly he showed a precocious talent for mathematics and astronomy, and he was elected a fellow of the college as early as 1794. Recognizing his promise, Cyril Jackson, by now Dean of Christ Church, became his friend and patron, and just as Robertson had deputized for John Smith, from around 1805 Rigaud gave lectures in astronomy and natural philosophy for the infirm Thomas Hornsby. In the same year he was elected as a fellow of the Royal Society, serving as a vice-president in 1837–38.

Rigaud's wife Christian, whom he had married in 1815, died in the same year that he was appointed to the Savilian chair of astronomy, and he was left with seven young children. According to his youngest son John, author of a privately printed *Memoir* of Rigaud, her death was 'an unspeakable loss' to his father, and only rarely could he bring himself to mention her. Oxford became the object of Rigaud's fierce attachment, and his ties to the city were so noteworthy that on his death the *Oxford Herald* reported that

> he had never been absent from Oxford so much as a single year during the period which has since elapsed, little short of half a century.

Rigaud's industriousness was also notable and he was apparently always at work, either in Oxford or Richmond, where following his father's death in 1814 he occasionally performed the role of official observer at the King's Observatory. His Oxford routine saw him spending the mornings on 'scientific and literary' activities, taking a walk in the afternoons before going to the Bodleian Library, with nights spent at the observatory.[38]

With the support of Jackson and others, Rigaud was encouraged to participate in University administration from an early point in his career, and he played a key role in the major overhaul of the examinations system that took place in the first decades of the 19th century. In 1800 the University created the honours degree, making the broader institution (rather than individual colleges) responsible for examinations. In 1807 Final examinations were classified and separated into two honours schools, Literae Humaniores and Mathematics and Physics. As Chinnery's letters make clear, one focus of the mathematical examination was the first book of Newton's *Principia*, but more broadly students were supposed to have a demonstrable knowledge of algebra and Euclidean geometry. Later in their degree, the students were exposed to

[*Left*] An oil painting by John Francis Rigaud of his nephew and niece, Stephen Peter Rigaud and his sister Mary Anne, in Richmond Park. [*Right*] A silhouette of Stephen Peter Rigaud.

the Newtonian version of calculus, but the analytical methods exemplified in the writings of Laplace and Lagrange would not filter into the University until the late 1820s, a decade after they had been introduced at Cambridge. Rigaud was a public examiner in mathematics in 1801 and 1806, and as Savilian professor of geometry he participated in the curriculum reforms of the 1820s.[39]

Rigaud was also made Reader in Experimental Philosophy, in succession to Hornsby, and for nearly three decades he gave 'abundant' lectures in the subject. In 1821 he published a synopsis of his course on the subject, reassuring students that the existence of such a booklet would help them to return more easily to lectures and to subjects that they had experienced earlier in the course. He added that in future he would make greater use of experiments, repeating them whenever he felt it necessary. Rigaud made copious use of schemes and models as teaching aids, and surviving correspondence with his machinist Alex Galloway (relating to the provision of model steam-engines for his classes) shows vividly how committed he was to the pedagogical benefits of using working machines as demonstration devices.[40]

Rigaud as observer, historian, and editor

As Radcliffe Observer, Rigaud was given an annual salary of £300, which was supplemented by the fees that he received for his lectures in experimental philosophy – for example, in 1819 students paid two guineas (£2.2s) to attend a term's worth of lectures in the subject. A list of the students attending his Savilian lectures in both geometry and astronomy from 1819 to 1838 reveals an average cohort of about twenty students per year over the entire period. The audience for geometry in any given term never dipped below 12 and was often in the low 20s, while attendance at the astronomy lectures Rigaud gave from 1827 was considerably lower, hovering between 6 and 15 students. This poor attendance lasted for four years, until Thomas Gaisford, dean of Christ Church, 'made the Ch.Ch. men attend', according to Rigaud's own annotation in

The meridian mural circle installed by Rigaud in the Radcliffe Observatory.

the register;[41] this requirement boosted the attendance considerably, with a peak of 39 students attending in 1836.

Throughout his tenure as Savilian professor of astronomy and Radcliffe Observer, Rigaud assiduously carried out observations with his assistant Angel Lockey, a task that entailed the concurrent recording of thermometric, barometric, and rainfall readings. But by the time that Rigaud took up his post, Bird's instruments – although still precise – were no longer the best available, while the quadrants that had put Oxford's Observatory at the forefront of late 18th-century observing had ceased to be used by any of the leading observatories. In 1836 Rigaud obtained a new 6-ft meridian mural circle with a 4.1-inch aperture, but it was only when this instrument became operational after his death that the Observatory could once again contribute substantively to contemporary astronomical observing.[42]

In December 1828 Rigaud and a number of his contemporaries, such as William Buckland, Charles Daubeny, and Baden Powell (Rigaud's successor as Savilian professor of geometry) founded the Ashmolean Society. This society, whose membership amounted to over 200 by 1835, was instituted to 'promote an interchange' of natural history, experimental philosophy, and other scientifically relevant topics. Rigaud contributed papers on topics such as meteorology and steam engines, but most of his presentations concerned historical subjects. At a time when Cambridge contemporaries such as John Herschel and William Whewell were writing major works on the history and philosophy of science, Rigaud was engaged in a typically Oxonian exercise, pioneering the in-depth study of the archival remains of British science.[43]

Rigaud began his career as a serious historian of science by producing an impressive edition of Bradley's papers, including a detailed biography of his subject. In his preface to the work, he noted that the task of completing it had been imposed on him, and he lamented that he had been

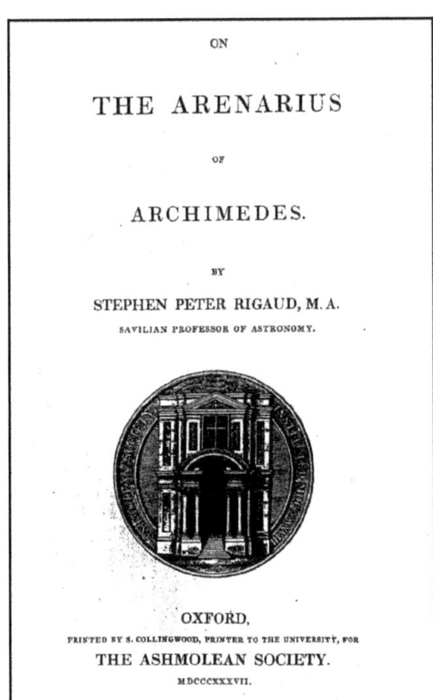
ON

THE ARENARIUS

OF

ARCHIMEDES.

BY

STEPHEN PETER RIGAUD, M. A.

SAVILIAN PROFESSOR OF ASTRONOMY.

OXFORD,

PRINTED BY S. COLLINGWOOD, PRINTER TO THE UNIVERSITY, FOR

THE ASHMOLEAN SOCIETY.

MDCCCXXXVII.

Rigaud's *On the Arenarius of Archimedes*, written while he was Savilian professor of astronomy.

forced to seek out many of the sources that he had used, wading through many boxes of largely disordered papers and drawing on the extensive collections of materials at the Bodleian Library and other repositories; these included papers at Greenwich and the records of observations made at Shirburn held by the Earl of Macclesfield. Observing that Bradley's data were still of great use, Rigaud noted that his studies and pursuits 'could be presented to the world in no way, which would make them more clear or more useful, than if they were connected with a narrative of his progress through life'. Although Rigaud's edition of Bradley's life and work has been superseded by recent studies on the subject, his commitment to acquiring information 'from original authorities' and his facility in handling the technical details of Bradley's work mark it out as an exemplary piece of scholarship.[44]

Once he had finished paying homage to Bradley, Rigaud devoted himself to publishing articles and longer works on topics as diverse as Halley's unpublished astronomical writings, a biographical account of the mathematician and instrument maker John Hadley, a history of the Radcliffe Observatory, Archimedes's *Arenarius*, an essay on the first publication of Newton's *Principia*, and the long-term development of steam engines. As we have seen, Rigaud was particularly drawn to the achievements of Thomas Harriot, and in a preliminary assessment of the latter's archive he remarked on Harriot's very early use of a telescope to discern the features of the Moon's surface, the satellites of Jupiter, and sunspots.

When Rigaud died in 1839 he had almost completed an edition of 17th-century scientific correspondence from the Macclesfield papers, including letters from Isaac Barrow, John Flamsteed, John Wallis, and Isaac Newton. It is a mark of his energy that he copied the entire mass of original correspondence into his own neat hand to make it ready for the press. Rigaud was concerned with the scions of science and also with its unsung heroes, and he made a serious effort to place individual achievements in their intellectual contexts. After his death a friend drew attention to Rigaud's 'peculiar delight in tracing the history of an invention, or of illustrating the biography of those who, however eminent in their day', had lapsed into obscurity.[45]

Conclusion

Thomas Hornsby, Abraham Robertson, and Stephen Peter Rigaud were the only occupants of the Savilian chair to combine their professorial positions with the role of Radcliffe Observer. As such, they were both university lecturers and positional astronomers, operating in institutional and intellectual contexts that differed considerably from those of such astronomers as their great contemporary William Herschel. Their working practices were variously constrained by the Savilian statutes, by increasing the demands associated with making astronomical observations, by their duties as Radcliffe observers, and by older Oxford traditions of scholarly editing. A combination of these factors meant that all three professors found it impossible to publish reliable data in print as regularly as other astronomers needed.

By the time that Rigaud occupied the position, Bird's instruments were nearly sixty years old. Although they were no longer cutting edge, they had provided an almost unbroken record of astronomical service from 1774 to 1838. The observations of Hornsby, Robertson, and Rigaud were all recorded in large folio books, in the same format as Bradley's had been. As the Radcliffe Observer A. A. Rambaut noted in 1900, 'while the results of their zealous labour in editing the latter have become the groundwork of accurate astronomy', their own observations – including around 130,000 transits and more than 60,000 zenith distances – remained unreduced and unpublished. Many of Hornsby's observations were published in the 1930s, but the bulk of their collective labours still remains unpublished and awaits serious scholarly study.[46]

Charles Pritchard (1808–93), Savilian professor of astronomy in the late 19th century, and founder of the University's 1875 observatory.

CHAPTER 6

The Victorians

ROGER HUTCHINS

For 54 years of Queen Victoria's reign, three men occupied Oxford's Savilian chair of astronomy: George Johnson (from 1839 to 1842), William Donkin (from 1842 to 1869), and Charles Pritchard (from 1870 to 1893). After losing the use of the Radcliffe Observatory, the professors found themselves constrained and buffeted by four influences: Tractarian challenges to the University, a revolution in astronomy as the Royal Observatory in Greenwich came to dominate meridian work while equatorial telescopes were becoming viable, continuing tensions between research and teaching, and the colleges' grudging responses to government commission enquiries into their finances and provision for science teaching. These pressures led inadvertently to the founding in 1875 of a major new observatory with research capability.

Introduction

In the mid-19th century Oxford University included twenty colleges and five small halls. Each college was a separate legal corporation, owned by its elected fellows with its own statutes, endowment, and income, and was typically founded for up to 30–40 fellows and a small number of scholars living within its walls. Even by the late 1860s the colleges were teaching fewer than 2000 young men in total.[1] Nearly all fellows were in Anglican holy orders and almost all were classics graduates. While respecting Cambridge's mathematical culture, Oxford believed that the Literae Humaniores classics curriculum was best suited for those destined for the expanding Church, politics, medicine, and non-technical government service.

The University was not wealthy. Its income derived from taxes paid by the colleges, which thereby had great influence. The increasing specialization of the sciences necessitated the provision of new buildings and teaching posts,[2] and international competition led directly to the instigation by the British government of the Russell and Devonshire commissions of enquiry in 1850 and 1871. Their recommendations engendered legislation in 1854 and 1877, consequent ordinances in 1857, and major statutory reform in 1882. Confronting a wall of collegiate vested interests, progress in science was glacial.

Roger Hutchins, *The Victorians.* In: *Oxford's Savilian Professors of Astronomy.* Edited by: Robin Wilson and Steven Balbus, Oxford University Press. © Oxford University Press (2025). DOI: 10.1093/oso/9780198894292.003.0006

The Tractarians

Between 1820 and the 1840s, developments in geology, disputes about the age of the Earth, and new discoveries of strange fossils challenged ideas of Creation and Design, with intellectuals arguing irreconcilable views. As urbanization rapidly increased, church attendance slumped and social unrest spread. To encourage Church influence, 600 new Anglican urban churches were built between 1824 and 1884, while the clergy grew in number from 14,500 in 1841 to 24,000 in 1875, thereby providing careers for Oxford graduates. The government begrudged reform, but abolished confessional 'tests' in 1828 and removed anti-Catholic discrimination in 1829.

Believing that scientists were compromising the Anglican integrity of the University and its governance, John Henry Newman and his traditionalist high-church friends precipitated the Oxford Movement of 1833, seeking Anglican institutional renewal by reverting to Roman Catholic doctrines and practices. A series of sermons and pamphlets (or 'tracts') by the 'Tractarians' polarized an increasingly vehement and personalized debate as they focused on the University's relationship with government ministers and on the most senior clerics' involvement with University appointments.

In 1835, intense Tractarian opposition defeated a campaign to admit dissenters to the University. In the following year, when the Prime Minister Viscount Melbourne nominated Renn Hampden for the Regius chair in divinity, a storm of Tractarian objections included characterizing Hampden as a heretic. This infuriated Melbourne, who insisted that the appointment was confirmed. Tempers were high, and so in 1839 the Savilian appointment needed to be swift and assertive.

New options in astronomy

Meanwhile, astronomy was serving the public needs for a governmental Nautical Almanac, essential for timekeeping and dependent upon the accurate location of the stars and planets. This 'useful and proper' work was required by navigators, geographers, map-makers, the military, and later the railways and telegraph. The observed positions of a target star had to be reduced by mathematical equations to its 'true' position. Such work was extremely laborious and time consuming, and publication was expensive.

This astrometry was the responsibility of the Royal Observatory at Greenwich, but by 1820 its inefficiencies had become a national scandal. The astronomer George Biddell Airy saw his opportunity for reform. In 1823 he had been Senior Wrangler and Smith's Prizeman at Cambridge, and in 1828 he was appointed to the mathematical Plumian chair of Astronomy and to the directorship of the Cambridge Observatory, recently completed in 1824. With Greenwich as his target, he devised a new system of reductions and published his results annually, to the astonishment of his contemporaries. For these reductions, Airy devised printed forms with standardized steps – his 'factory method', implemented by supernumerary computers and with teenagers working up to twelve hours a day. Creating such a model institution at Cambridge made him the ideal candidate to succeed as Astronomer Royal in 1835, and to drive this work he appointed the Revd Robert Main, sixth wrangler in 1834, as his First Assistant.

Meanwhile, in 1832, Airy's report to the British Association Meeting in Oxford had concluded that at Oxford's Radcliffe Observatory, 'grinding the meridian was going on steadily

The new University Museum of 1860.

and perseveringly, without the slightest thought of reduction and publication'.[3] But astronomical research involved risks. In 1835 the Radcliffe Observatory acquired a new Jones 4¼-inch mural circle with a unique design for £350, but by 1841 it had already cost £1081. The next renewal in 1849 was a small observatory with a 7½-inch heliometer, and with its building it cost £3345.[4] Such unanticipated costs for new technology were a salutary warning. New specializations evolved in astronomy, but the unendowed university observatories had no academic need for them, and they could not afford to be risk-takers in such research pursuits.

Teaching and research

The Cambridge Observatory of 1824 supported research-led teaching for its most gifted students, and its University's modernized curriculum attracted the nation's mathematical talent. But in Oxford the mere existence of the Radcliffe Observatory (with no college engagement) could not motivate Classics dons to revise the curriculum. In 1849–50, anticipating the Russell commission, an easy deflection was to establish new schools of law and modern history, and the natural sciences. For the last of these a splendid new museum was completed in 1860 to house the teaching collections of its six professors. The two Savilian professors were each given a room and a shared lecture room, but astronomy was not included in the new school.

The student pool was tiny. Even as late as 1913, parliamentary statistics recorded only 5.8 per cent of 14- to 16-year-old boys in England and Wales being in education, with all in fee-paying grammar and public schools and with fewer than 5 per cent entering universities. This was an impediment to reform. In 1872 Oxford had four times as many scholarships in classics as in mathematics, and the elite schools trained their students to gain those scholarships.[5]

Russell's commission of 1850 required reform, and Devonshire's commission of 1879 recommended it, but the colleges resisted the University's attempt to gain power by building common science laboratories for new subjects, staffed by university employees paid for by college taxes. As the historian Richard Brown has summarized:[6]

> The watershed for Oxford and Cambridge came after the Cleveland Commission of 1873 led to the Universities of Oxford and Cambridge Act of 1877 and the consequent revision of the statutes of colleges in 1882. They were obliged to release some of their funds for the creation of scientific professorships and university institutions. Only then, with this rebalancing of power between colleges and the universities, was it possible to create an Oxford and Cambridge more oriented to research in science and scholarship, professional training, a widening curriculum and a strong professoriate.

A rare success for Baden Powell, the Savilian professor of geometry, occurred in 1831 when he secured funding for three University mathematics scholarships of £50 per year for three years, later amended to four awards of £30 for two years, to fund postgraduate study – at last there was a pathway to careers. Early holders were William Donkin who later became the Savilian professor of astronomy, Bartholomew Price who became the Sedleian professor of natural philosophy (the University's third mathematical chair), and Henry J. S. Smith who was appointed to the Savilian chair of geometry in 1861 following the death of Baden Powell.

But the Victorian mathematics professors had no leverage. Until 1887 all undergraduates spent their entire first year on Literae Humaniores, and it was only in 1928 that Oxford forged a formal link between analytical mathematics, physics, and astronomy.

George Johnson

Born in 1808, George Henry Sacheverall Johnson was the third son of a clergyman in Keswick, Cumberland. He matriculated in Oxford at age 17 at The Queen's College in May 1825, winning a college scholarship in 1826, and being awarded the University's new Ireland scholarship 'for the promotion of classical learning' in 1827. Gaining a double first-class degree in Literae Humaniores and mathematics in 1828, he graduated with a BA degree in 1829. On 25 June 1829 he was elected a Taberdar, a status of College preferment and commitment for a candidate awaiting a fellowship vacancy. Having received his MA degree in 1833 and ordained a deacon in 1834, he was appointed college chaplain in 1835 to tide him over. Meanwhile, in 1831, he had been the first to win a new University mathematical scholarship. He served as the University's mathematical examiner in 1834 and 1835, and later from 1850 to 1852.[7]

George Johnson produced two scientific papers while he was a tutor. In 1835 Baden Powell submitted to the Royal Society Johnson's 'A new method of discovering the equations of caustics' (a subject within optics) on his behalf; the Royal Society declined to publish it in full, but published an abstract in 1837. So in 1835 Johnson had his 33-page treatise *Optical Investigation: Caustics* privately published in Oxford for 'The Mathematical Society' a self-identifying subgroup of Ashmolean Society members. These papers supported his election to a Fellowship of the Royal Society in January 1838.

The 1839 election

After Thomas Hornsby's death in 1810, it seemed natural for the University to transfer Abraham Robertson, his co-editor of Bradley's observations, from the Savilian geometry chair to astronomy, and in 1827 for Stephen Peter Rigaud, an experienced observer, to be similarly moved (see Chapter 5). These University initiatives were followed by the Trustees accepting each incumbent as the Radcliffe Observer.

However, the Trustees needed their Observatory to work effectively. In 1811 they had provided for an observing assistant, and in the following year they doubled Robertson's inadequate Savilian stipend of £150 on condition that he would prioritize observations and their publication. Unfortunately, these conditions introduced tensions between the divided responsibilities of the Observer and the professor paid to teach, and in 1834 they accepted Rigaud's plea for instrument renewal.

As we have seen, for the election of the Regius chair of divinity in 1836 Lord Melbourne insisted upon his prerogative. In 1839 he was still Prime Minister, and the 'election' to succeed Rigaud as Savilian professor of astronomy can be understood only within the context of the Tractarian dispute. Of Savile's nine 'most distinguished' electors, six were judges, two were bishops, the University's Chancellor was the Duke of Wellington, and the only Oxford graduate was the Chairman, Archbishop William Howley.

Well aware of Tractarian vehemence and of Melbourne's views, these appointed office holders sought the opinion of Ashurst Gilbert, the resident Vice-Chancellor and Principal of Brasenose College, who was of high-church opinion and averse to Catholic ceremonial. In order to maintain Anglican broad-church policies and prerogatives, they would not be concerned with the workings of the privately owned Radcliffe Observatory. Furthermore, in George Johnson the Vice-Chancellor had a recommendable candidate: an Oxford-made academic, a conservative Anglican, and the outstanding available mathematician who in 1837 had lobbied Convocation against a revision of University statutes, arguing that it could not initiate policy.[8] These qualities outweighed his lack of astronomical knowledge or inclination.

However, for the politician Robert Peel, a Radcliffe Trustee since 1828 and an Oxford contemporary of Gilbert, an experienced astronomer was the crucial criterion. The once-finest observatory in Europe had slipped to obsolescence, with its work derided in 1832, so that Peel had supported Rigaud's 1834 appeal for instrument renewal. The University's carelessness confronted Peel with the immediate options of prioritizing the Trust's investment by appointing a competent Observer or acquiescing to the fait accompli of Oxford's compromised politics.

Peel lost no time in consulting John Herschel, president-elect of the Royal Astronomical Society, who was about to take office in May and who knew an individual who was shortly to graduate and who aspired to be a professional astronomer. Between 1829 and 1833, while a lieutenant in the East India Company on St Helena, Manuel Johnson had observed a catalogue of 606 southern stars. In early 1835 his catalogue was published, and he was promptly awarded the Royal Astronomical Society's Gold Medal.

At age 30 Manuel Johnson matriculated from Magdalen Hall in December 1835, graduating with a Literae Humaniores BA degree in 1839. But with neither a mathematical degree nor an MA degree, he was not eligible for the Savilian chair. Moreover, 1839 was the peak of the Tractarian agitation, and 'Manuel Johnson was a . . . most convinced follower of the Movement to the day of his death'.[9]

In a Congregation, holden on Thursday last, the

The Trustees of the Savilian Professorship of Astronomy have elected the Rev. George Henry Sacheverell Johnson, M.A. Scholar and Tutor of Queen's college, to the vacant chair in this University.

The Trustees to the will of Dr. Radcliffe have appointed Manuel John Johnson, Esq. B.A. of Magd. hall, to the office of Radcliffe Observer. Besides the salary, there is a house for the Observer, apartments for observation and lectures, as well as rooms for an assistant observer. We are informed that Mr. Johnson is indebted for his preferment to the recommendation of Sir J. Herschel, Bart.

An announcement of the separate but coincident appointments of George Johnson as Savilian professor and Manuel Johnson as Radcliffe Observer.

The crucial sequence of who caused the schism between the University and the Radcliffe Trustees was as follows.[10] In March 1839 the 30-year-old George Johnson had found himself the University's choice for the Savilian chair. On 25 April, the day before the Trustees first met, the electors appointed him for the Savilian chair. The Trust then recorded their appointment of Manuel Johnson as Radcliffe Observer on 1 May, the day before Convocation confirmed George Johnson in the Savilian chair. The electors' haste and neglect of the Trustees' views broke Oxford's long tradition of the teaching of theoretical and practical astronomy by an active practitioner. As Ivor Guest, secretary to (and historian of) the Radcliffe Trust, explained:[11]

> When Robertson and Rigaud after him were appointed, not only was a salary paid but conditions of employment were carefully prepared so that there could be no doubt where the authority lay . . . but now . . . The University, it seems, had omitted to consult them when they appointed George Johnson to the Savilian chair . . . Their own soundings . . . produced at least two names, for on Friday, 26 April 1839 they met to consider the applications of the several Candidates . . . and adjourned to give the matter further consideration on the following Wednesday, 1 May, when they appointed a brilliant young astronomer, Manuel Johnson. Such an act of independence must have come as a profound shock to the University, which was deprived at a stroke of the use of the Observatory.

Both appointments were announced in the 'University and Clerical Intelligence' column in the *Oxford City Chronicle* of 4 May. Over the next twenty years Manuel Johnson proved highly effective as Radcliffe Observer, restoring the Observatory's utility and reputation, and creating the base for a century of Radcliffe astronomy in Oxford and later in South Africa.

Because George Johnson's appointment had been confirmed on 2 May, the day after the Trustees had appointed a different Radcliffe Observer, he lost the facilities of that post: a desirable house that would have enabled him to seek a wife, and the Trustees' stipend of £150.

These circumstances were not of Johnson's making, but the changes would undoubtedly have come as a great shock. In consequence, he cobbled together a college living and fees. In 1839 The Queen's College had permitted him to assume the Savilian chair without losing his college positions, but when he was duly proposed as a Fellow in March 1842, the Savilian statutes prevented him from holding college offices. He chose to accept the fellowship and resigned the Savilian chair in that month.

Then his luck improved. Shortly after Johnson had gained his college fellowship, the Revd Charles Stocker resigned the White chair of moral philosophy after just one year, and took a parish. Although the stipend of £100 was too meagre to support the chair, it could be held while retaining his college fellowship. Johnson sought it, and by December 1842 he had been successful. His fellowship stipend of about £280, together with the fees from his college offices, averaged about £320 per year. He received the stipend and attended Governing Body meetings regularly until 1855.[12]

George Johnson would have been justified in feeling that the University had used him carelessly. Pursuing college and church careers was a clear choice away from science and mathematics, while moral philosophy remained clerical and classical. Its tenure strengthened his qualification for clerical preferment, but he resigned the post in 1845. The 1851 census found Johnson still living in The Queen's College as 'Clergyman, Fellow and Tutor'. As the University historian Mark Curthoys has observed:[13]

> Both elections took place at the height of the Tractarian movement, and Johnson's known strong antipathy to it may have influenced each appointment.

Johnson's lack of publications suggests that he was not at heart a scientist. He occupied the Savilian chair, but had no discernible interest in astronomy. The only incumbent not to have sought fellowship of the Royal Astronomical Society, he had no intention of representing his University there. Johnson's tenure was surely a low point for Savile's chair of astronomy.

Johnson's later life

Because the appointments of George Johnson and his two successors reflect the Tractarian pressures, we should understand the networks and prerogatives of appointment that determined his subsequent career.

Johnson was one of the leading Oxford tutors of his day, and among his private pupils he befriended Archibald Tait who became Bishop of London in 1856 and Archbishop of Canterbury in 1868. Another friend was Arthur Stanley who became Dean of Westminster.

He was also a reformer who allied with Stanley and Benjamin Jowett in pressing for a royal commission to enquire into the 'state, discipline, studies and revenues' of the University and colleges, and was then appointed by government to the Russell Commission of 1850. Johnson's friends Tait and Baden Powell were also commissioners, and Stanley had the influential role of secretary, drafting the report and its recommendations. Johnson suspected that his service as a government-appointed commissioner was unpopular in the University and had cost him becoming the Provost of Queen's in 1855 (and often a college headship led to a bishopric). Yet because he was 'one of the university reformers whom the clerical party found least objectionable . . . on this account he was made one of the executive commissioners under the 1854 Oxford University Act to revise the statutes of the University and colleges'.[14]

The Church.

THE DEANERY OF WELLS has been conferred by the Earl of Aberdeen on the Rev. George Henry Sacheverell Johnson, M.A., Dean and Fellow of Queen's College, and one of the members of the late Oxford University Commission.

The conferral of the Deanery of Wells to George Johnson in March 1854.

Significantly, for the three years 1852–54 the Bishop of London selected Johnson as Oxford's 'Whitehall preacher' in the Chapel Royal at St James's Palace. These appointments, for one Oxford and one Cambridge man, were considered a reward for distinguished service in the universities, and often led to promotion. The chapel congregations comprised the royal family together with ministers of government, statesmen, and literary men.

Johnson was rewarded for his services to the Russell commission by being appointed Dean of Wells Cathedral, an office widely reported to be worth £1500 a year. This appointment was a prerogative of the Prime Minister, William Gladstone, and was made by the Earl of Aberdeen on 10 March within just four days of the death of the incumbent. Five weeks later, in April 1854 at age 46, Johnson married Admiral Robert O'Brian's 29 year-old daughter Lucy.

In 1855 Johnson relinquished his Oxford fellowship and stipend. His incumbency of Wells was marred by long-running criticism from 'The Enemies of the Dean' after he had appropriated the vicarage of St Cuthbert's parish in Wells, thereby gaining its annual £310 stipend. The influential 'Enemies' petitioned the Prime Minister against the 'evils' ensuing from this plurality. Johnson defended himself in the press, but resigned his St Cuthbert's position in 1870, and thereafter he was rarely seen:[15] 'For some years . . . in great retirement and many periods of absence . . . in part due to long declining health, a sensitive temperament, and reserved habit'.

Johnson published three religious works. Following a decade of poor health, he died at Weston-super-Mare on 4 November 1881, aged 74 years. The *Church Times* commented unenthusiastically on his tenure:[16]

> George Johnson in 1839 was the most credible candidate at a fraught time. In the University he later made his mark by good service on two government commissions, then gained his reward.

William Donkin

When Johnson resigned Savile's chair of astronomy in December 1842, a more impressive young scholar, William Fishburn Donkin, was encouraged to apply. He was born in February 1814 at Bishop Burton, Yorkshire, the fourth of eleven children of Thomas Donkin, a land

George Johnson, Dean of Wells Cathedral, in 1861.

agent, and his wife Alice Bateman of Whitby. Thomas's brother was a leading engineer and a fellow of both the Royal Society and the Royal Astronomical Society, while Alice's brother was a leading doctor.

William became the eldest surviving son, and when he was 14 the family moved to York so that he could attend the famous St Peter's School. During his three years there he gained a love of classics. In October 1852 he matriculated at St Edmund Hall in Oxford, and in December 1834 he migrated to University College on a classical scholarship, and became the first student there to achieve a 'double first' in classics and mathematics. He graduated with a BA degree on 26 May 1836, was elected to a fellowship on 3 December 1836, and was admitted as a fellow six months later. In 1837 he was awarded the University's senior mathematical scholarship and the Johnson mathematical scholarship, and was appointed University College's praelector in mathematics. These dazzling achievements confirmed his reputation in the University, and he was 'the first Fellow of the College to devote himself exclusively to the sciences'.[17]

Donkin's MA degree was conferred in April 1839, but he did not take holy orders. He continued lecturing at St Edmund Hall from 1836 to 1842, while also lecturing and tutoring at University College. He was Public Examiner in mathematics and physics in 1840, 1845, and 1853, and examiner for natural science in 1855.

Donkin was a member of the Ashmolean Society, but had not yet published any research papers when, at the age of only 28, he was 'advised to stand for the Professorship of Astronomy' following George Johnson's resignation in March 1842.[18] Donkin's proposer may have been Frederick Plumptre, the Master of University College since 1836.

William Donkin around the time of his appointment as Savilian professor of astronomy.

Donkin was duly elected on 3 May 1842. After a year's grace he resigned his college fellowship on 6 May 1843. In January 1842 he had been elected a Fellow of the Royal Society, and in November of that year was elected a Fellow of the Royal Astronomical Society.

In June 1844 in St Peter's Church, Oxford, William Donkin was married at the age of 30 to Harriet Hawtrey, the daughter of a clergyman. They moved into the professor's house in New College Lane, where Harriet produced a daughter and four sons.

Donkin was shy but cheerful, good humoured, and played the piano and the violin; soon, William and Harriet's musical evenings at home became a social centre within the university. He loved the classics, with a particular interest in analysing music and acoustics, and in the 1840s he wrote several articles on ancient Greek music. From 1842 he lectured on physical astronomy, and after 1844 if students sought practical lessons, Halley's observatory accommodated measurements with hand-held instruments.

By 1842 Donkin had already demonstrated his brilliant mathematical skills. There was no thought of replacing the recent loss of the Observatory, and nor could Donkin and Baden Powell conjure a student clientele. But through the range of their published researches and their beneficial participation in the Ashmolean Society as it influenced the seniors of Oxford, they put 'Oxford mathematics back on the national map'.[19] Indeed, between 1844 and 1864 Donkin published twenty mathematical papers,[20] contributing to statistics, solving Laplace's equation in areas of physics, extending Hamiltonian mechanics, and writing twice on differential equations. Donkin's obituarist for the University and for the Royal Astronomical Society is believed to have been his friend Bartholomew Price, the Sedleian professor of natural philosophy, who recalled:[21]

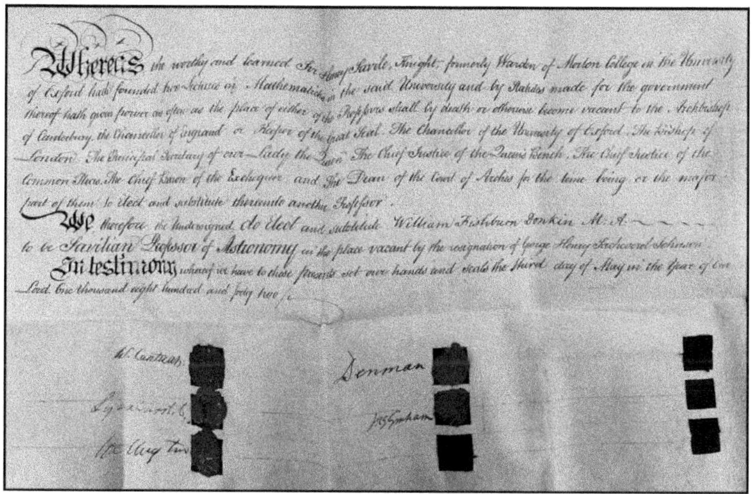

Whereas the worthy and learned Sir Henry Savile, Knight, formerly Warden of Merton College in the University of Oxford hath founded two Lectures in Mathematics in the said University and by Statutes made for the government thereof hath given power as often as the place of either of the Professors shall by death or otherwise become vacant to the Archbishop of Canterbury, the Chancellor of England or Keeper of the great Seal, The Chancellor of the University of Oxford, The Bishop of London, The Principal Secretary of our Lady the Queen, The Chief Justice of the Queen's Bench, The Chief Justice of the Common Pleas, The Chief Baron of the Exchequer and The Dean of the Court of Arches for the time being or the major part of them to elect and substitute thereunto another Professor.

We therefore the undersigned do elect and substitute William Fishburn Donkin M.A. to be Savilian Professor of Astronomy in the place vacant by the resignation of George Henry Sacheverel Johnson.

In testimony whereof we have to these presents set our hands and seals the third day of May in the Year of our Lord One thousand eight hundred and forty two.

W. Cantuar

Denman

Donkin's election in May 1842; the Electors identified here (including the Duke of Wellington) illustrate the political and church grip upon patronage for University appointments.

> He was a mathematician of great attainments ... taking an especial interest in the investigation and application of the higher theorems of analysis.

It was of less importance that Donkin's research interests had not been in astronomy than that he stuck the course for 26 years, as by his personal influence and his publishing he restored the reputation of the chair. However, with instruments of his own he taught practical astronomy, and restored good relations with the Radcliffe Observers.

Another particular issue engaged him. In 1853 John Couch Adams had given the Royal Society his landmark paper on the secular acceleration of the Moon's mean motion, a vital part of his lunar theory for which he was later awarded the Royal Astronomical Society's Gold Medal. For more than a decade a violent controversy ensued between French mathematicians and those from Britain (Lubbock, Cayley, and Donkin). In 1861 Donkin read his only paper to the Society, on the Moon's motion, in which he validated Adams's result.[22]

New College and the Savilian stipends

Following the Russell Commission's report of 1852 and the Reform Act of 1854, the Ordinances of June 1857 imposed obligations upon particular colleges to open restricted foundation funds to pay stipends to designated professors whose chairs would be annexed to the college. Responses to this proposal varied.

The Savilian chairs were duly attached to New College, whose original foundation in 1369 had been for '70 Fellows and Scholars', all from Winchester College and all resident within New College. Following the Commissioners' aim to open half of those 'closed' fellowships and studentships to competition, the 1857 Ordinances imposed a new constitution for an eventual maximum of forty fellows by examination, to follow a natural reduction to thirty fellows by

In 1841 the Radcliffe Trustees commissioned this magnificent 7½-inch Merz–Repsold heliometer from Hamburg. When it finally became operational in 1848 it had the potential for making the Radcliffe Observatory one of the best equipped observatories in the world for precision astrometric work, but unfortunately problems with its innovative divided-glass lens frustrated these hopes.

death or resignation. At that stage the Savilian professors' annual salary, initially of £150 each for five years and thereafter £300, 'shall become payable'.[23]

A simultaneous debate at The Queen's College in 1856–57 followed the election in 1853 of Bartholomew Price to the Sedleian chair of natural philosophy. By 1858 the College fellows had agreed to contribute to the Sedleian stipend, and 'a new professorial statute was approved by royal assent in April 1858',[24] with a previously variable annual stipend now guaranteed at £300, to be made up from the University Chest. In particular, the new Sedleian statutes included the important precedent that its electors should include the two Savilian professors, the Astronomer Royal, and the President of the Royal Society, so as to ensure the scientific credentials of the candidates. That enhancement was not achieved for the Savilian chairs until 1882, even though the Act of 1854 had conferred upon the Radcliffe Observer the status of a professor within the University.[25]

Meanwhile in New College the Warden's minutes for 29 April 1856 had recorded a similar debate, but with much obduracy:[26]

> It was agreed that any diminution of the number of the Foundation members with a view to the endowment of the University [Savilian] Professoriate would be prejudicial to the College.

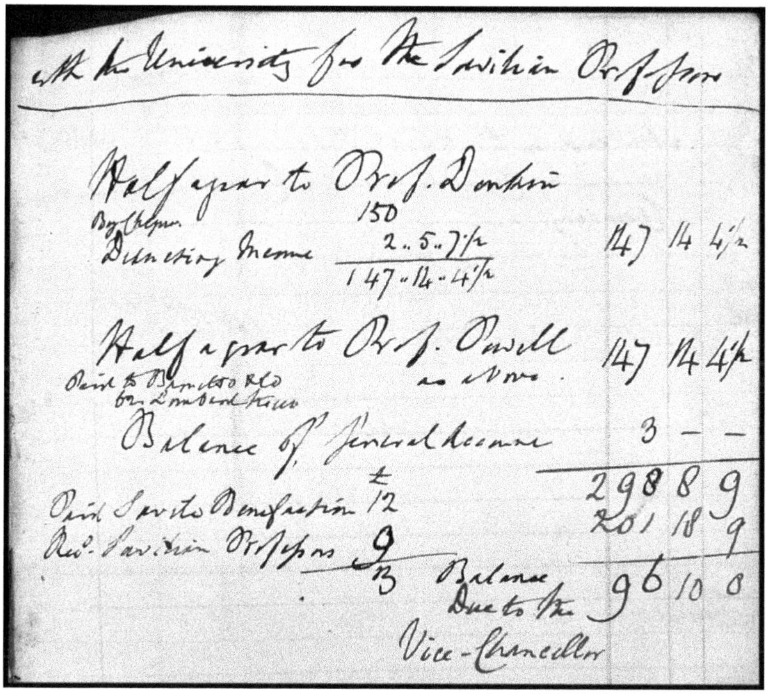
...with the University for the Savilian Professors

Half year to Prof. Donkin
By Cheque 150
Deducting Income 2..5..7½
147..14..4½ — 147 14 4½

Half a year to Prof. Powell as above. 147 14 4½

Balance of General Account 3 – –

298 8 9
Paid Savile Benefaction £12
Recd. Savilian Professors 9 — 201 18 9
3 Balance Due to the Vice-Chancellor 96 10 0

William Donkin's stipend in 1858.

Vigorous discussion continued, but even by 1868 New College's Queen's Counsel confirmed that 'No payment has yet been made to the Savilian Professors'.[27] Meanwhile, the University's new 'Accounts Book for the Savilian professors, 1858–76' records that, following The Queen's College's support for the Sedleian chair, the University Chest took on the topping-up of the Savilian Endowment stipends to £300, as required by the Ordinances. Although the Savilian Estates Account's income varied, in 1865 and 1868 a stipend of £400 was achieved for each chair.

For William Donkin, the University's commitment in 1858 to cover New College's prevarication in providing his Savilian stipend provided assurance. From 1844 to 1857 his income had been the Savilian stipend of £150, plus the University College's lectureship of £170. But the changes for 1858 guaranteed a minimum of £300 from the University, which enabled Donkin to resign his lectureship. These college debates and the consequent resorts to University expedience illustrate the University's difficulty in complying with the Russell Commission's ordinances, and the modernization of restraints within the history of the Savilian chairs. On 26 March 1858 Donkin was elected as the first ever honorary fellow of University College, testifying to the esteem in which he was held at that college.

The death of William Donkin in 1869 finally precipitated a solution. Although New College had been designated in 1857 as the home of the Savilian professors this had not yet happened, and the college had paid them nothing. However, the Warden was now one of the electors, and when Donkin's successor Charles Pritchard, not being an Oxford graduate, required an MA degree to be allowed to teach, he was immediately invited by the Warden to matriculate there upon his election in February 1870. His first year's stipend was also voluntarily topped-up

[*Left*] The University Museum Keeper's house. [*Right*] The Museum Observatory of 1860–75, on a restricted site behind the Museum.

by New College, who paid him a further £147.10s 'for 1 year to Lady Day' [25 March 1871].[28] Thus, after fifteen years, the College's first steps towards fully funding the Savilian stipends were finally being amicably resolved.

Recombining offices

William Donkin had been one of the University's representatives to the Russell Commission's report on the provision for science. Their enquiries stimulated the University to complete its Museum in 1860. Donkin sought only a small teaching observatory.

In 1854 New College had revoked its leases for the two houses in New College Lane, and Donkin moved to 34 Broad Street. He would then be without an observatory to teach in, until the little Museum Observatory was completed at a cost of £167.5s. Manuel Johnson filled that gap. After 1860 Johnson's successor was Robert Main, aged 52, who for twenty-five years had been the First Assistant at the Royal Observatory in Greenwich. Main sometimes presented Donkin's lectures for him, and gave practical instruction when required, until his successor Charles Pritchard achieved his new observatory in 1875.

Meanwhile, Manuel Johnson's death in 1859 had created a unique opportunity for recombining the two posts. George Bramwell, Secretary to the Radcliffe Trust, hoped to reduce the Observer's salary of £400, but Donkin believed that 'Academic Teaching was incompatible with directing a Public Observatory'.[29] The University took the matter seriously, while Dr Henry Acland and John Phillips (about to take up the post of Keeper of the Museum) made enquiries. Phillips urged reunification, hoping that equatorial work might stimulate Natural Science students, while George Airy and Professor Thomas Robinson in Ireland concurred with Donkin on separation.

Donkin had the final veto, and his decision set the University and the Radcliffe's astronomy on divergent paths. Preserving the independence of the Radcliffe Observership was crucial to astronomy's career structure. The leading candidate for this position was Robert Main: continuously on the Royal Astronomical Society's Council since 1841 and a Gold Medallist in 1858, he was currently president from 1859 to 1861. A prodigiously productive astronomer who had longed for an independent post, Main moved into the Observer's house on 1 October 1860.

The interior of Oxford's University Museum. Constructed in brick and iron, the Museum realized Oxford's mid-century aspirations to improve the facilities for teaching mathematics and science in terms that recalled the soaring confidence of its medieval buildings. It included a teaching room upstairs for the Savilian professors of astronomy and geometry to share, and the small teaching observatory requested by Donkin.

Main's was another first-class mind, and he respected and shared Donkin's view that teaching undergraduates was incompatible with research. In 1863, perhaps stimulated by involvement with Donkin, he published *Practical and Spherical Astronomy*. His friend and obituarist Charles Pritchard stated in 1878 that it was 'the only book extant written by a real astronomer from which the student can learn real astronomy'.[30] By Donkin's 1859 rejection of recombination, and by the precedent of the University's paying for the Museum Observatory, its obvious inadequacies in 1870 provided a lever which, when exploited, would lead to the creation of the University's Observatory of 1875.

William Donkin around 1868.

Donkin's final years

From 1854 Donkin suffered an early onset of phthisis, a wasting disease like tuberculosis. This began to impose periods of rest, and he spent some parts of the winter in warmer climates. In 1857, at age 43, with a house full of small children, and with his health causing concern, his increased stipend had enabled him to resign his lectureship.

Donkin's health declined during the following decade, but through Main's friendship and involvement a knowledge of Oxford's situation became recognized by the Royal Astronomical Society's Council, not least by Warren De la Rue, its secretary from 1855 to 1863, and by Charles Pritchard, its secretary from 1862 to 1866 and president from 1866 to 1868.

Donkin aspired to combine his love of mathematics and music in a *magnum opus* on theoretical and practical acoustics. In 1867 he started to work on it, but was continually interrupted by severe illness and had to spend the winters of 1867 and 1868 abroad. He was working until three days before his death at home on 15 November 1869 at the age of 55. Bartholomew Price promptly brought the first part of Donkin's *Acoustics* to publication in early 1870.

William Donkin's death was greatly lamented in the University. As a remarkable tribute, twenty-six years after he had relinquished his fellowship, his colleagues memorialized him with a plaque in Latin in the chapel of University College:

> Most learned professor, expert in the heavens, music, religion, harmony (if anyone is). He was of a meek nature, beloved of his own, a most delightful mind glowing with extraordinary brightness, brilliant, high-minded, sharp, most wise, capable of all things, if only the powers of his body and mind might have equalled his indefatigable strength.

DEATH OF THE SAVILIAN PROFESSOR OF ASTRONOMY.

Mr. William Fishburn Donkin, M.A., Professor of Astronomy, Honorary Fellow of University College, F.R.S., died on Tuesday morning, after a lingering illness, at his residence, Broad-street, Oxford. Mr. Donkin succeeded Mr. George Henry Sacheveral Johnson, M.A., of Queen's College, to the Professorship in 1842. The endowment of the Professorship is £400 per year, and is open to persons of every nation, provided they are of good reputation, eminently well versed in mathematics, have a tolerable knowledge of Greek, and are 26 years of age. If they are Englishmen they must be of the degree of Master of Arts at least. The electors are the Archbishop of Canterbury, the Lord High Chancellor of Great Britian, the Chancellor of the University, the Bishop of London, the Secretary of State for the Home Department, the two Chief Justices, the Chief Baron of the Exchequer, the Dean of the Arches, and the Warden of New College, taking into their counsel the Vice-Chancellor of the University. The illustrious persons are solemnly conjured by the founder to seek for the ablest mathematician in other countries as well as our own, and without regard to particular universities or nations, to elect those whom they shall deem best qualified for the office. On a transmission of their choice the person so elected is admitted by the University in congregation.

Professor Donkin's death in November 1869 was reported in the London and provincial newspapers.

Worn out by a long illness, he died, prematurely as far as we are concerned. Most dear to his friends, heeded by all, much missed by the University.

Seeking Donkin's successor

By this time, thirty years on from George Johnson, a new class of men with industrial wealth had become the 'Grand Amateur' risk-takers who grasped the new photographic, spectroscopic, and visual opportunities with large apertures, and established the foundations of astrophysics. Upon their fertile ground the academic astronomers would need to compete.

In 1872 George Airy pronounced excoriating views on the five British university observatories, all of which were struggling. In particular, he criticised Oxford's Radcliffe Observatory for lacking the potential to engage with 'astronomical physics', claiming that Robert Main's meridian observations were 'of no use in the world'.[31]

The government's laissez-faire policy provoked the British Association for the Advancement of Science to advocate state endowment for scientific research.[32] This led in 1870 to the Devonshire Royal Commission on Science Instruction and the Advancement of Science, with three reports putting pressure on Oxford and Cambridge universities. Warren De la Rue put the case for major chemistry and physical laboratories, and for state support for the cultivation of the physics of astronomy. His disapproval of unseemly competition for government funding led De la Rue to resign from the Council of the Royal Astronomical Society in November 1872, and at its Annual General Meeting in February 1873 Pritchard (as a Council member) became so vehement that it was reported that:[33]

It was certainly a matter for regret that a learned and reverend Professor holding office in the Society should have allowed his feelings to get the better of his manners.

Thus, during the 1870–73 cusp in recognition of the new astronomical physics, coincident with his taking up the Savilian chair, Pritchard was fully engaged in the Astronomical Society's affairs and in working to achieve new infrastructure at national level.

In 1880 a proposed BSc degree in Natural Science was rejected as being too vocational, because it contained no Greek. In 1872 the twenty colleges had offered 55 scholarships in classics, 13 in mathematics, and only 6 in natural science,[34] and so the Devonshire Commission's Third Report of 1873 again recommended the appropriation of college endowments to fund science lecturers or professors, as well as proposing pensions for professors, science laboratories, more scholarships, and the establishment of Doctor of Science degrees based on original research.[35] These were all embodied in the Act of 1877, and were enacted in Oxford's New Statutes of 1882.

Oxford's mathematics professors remained hamstrung. In 1883 James Sylvester assumed the Savilian geometry chair, but found the teaching of Euclid's geometry to be 'miserable'. His response was in 1888 to launch the Oxford Mathematical Society, of which Pritchard and the Radcliffe Observer were founder members.[36]

In the meantime, while Donkin's health was declining, a major benefaction had set a precedent. In 1867 Robert B. Clifton, a sixth wrangler and Smith's Prizeman from St John's College, Cambridge, was appointed to Oxford's chair of experimental philosophy. In May of that year the Clarendon Trust offered £10,000 to build a department for him, but even as Clifton was in discussions with the architects, 120 members of Congregation urged instead that the money be spent on a swimming pool, while another group lobbied for a riding stable in the park. Fortunately, wisdom prevailed, and the new Clarendon Laboratory for physics was opened in October 1873, the first purpose-built laboratory in England. The cabinet containing physical apparatus was extremely antiquated, and the University granted Clifton £150 a year for three years to make arrears, and an annual budget of just £90.

Clifton produced his only research paper in fifty years, before settling to his own 'mission teaching young persons to become teachers'. Undergraduates had the apparatus for weighing and measuring, but Clifton did not encourage research. In 1870 mechanics and physics became an option in the Moderations examinations, and the new statutes of 1882 that provided for new Moderations in mathematics conceded the option for Honours finalists to take one subject of physics, chemistry, or biology within the Natural Science school. That status, still denied to astronomy in either school, brought Clifton an increasing number of students.[37]

Charles Pritchard

Charles Pritchard was born in Shropshire in February 1808, but in that year the family moved to Brixton in south London. At age 12 he lost his mother. His father was a hatter whose business was precarious, but the family helped. At John Stock's Academy the curriculum inspired Pritchard with its experimental practical learning, and he learned to use surveying and astronomical instruments, scales, and compasses. Described as a very gifted and highly excitable young man with restless energy but a nervous temperament, he had a good sense of humour, a kind nature, and was a self-improver. Financial difficulties disrupted his education, and for two years he usually worked alone, mainly on mathematics.

Helped again by family and friends Pritchard entered St John's College, Cambridge, at the age of 17 in 1826, was fourth Wrangler in 1830, and in 1832 was elected to a fellowship which gave him independence. Although offered a tutorship at St John's, he left there in 1833 to become the headmaster of a grammar school in Stockwell in south London, because this would enable him to marry. He married Emily Newton in December 1834, and they produced two sons and five daughters of whom two daughters died as infants. Emily died in February 1857, and at age 50 in 1858 Pritchard married Rosalind Campbell, twenty-three years his junior, who had a modest fortune which enabled them to retire in 1867 to the Isle of Wight. She bore him three sons and five daughters, including Ada who compiled and published a careful assessment of Pritchard's Oxford years by his Savilian successor Herbert Hall Turner.[38]

Fired by the exciting developments in science and technology that had taken place since the 1830s, Pritchard's curriculum at Stockwell was 'A systematic course of instruction relating to physical phenomena', but a disagreement over his refusal to use the Eton Latin grammar led the governors to forbid him from teaching science, and he resigned. The school parents, seeking a modern curriculum, then founded Clapham Grammar School, and he became its headmaster from 1834 to 1862. For boys aged from 12 to 16 he taught Latin and Greek, but he packed the school with scientific apparatus and an observatory in 1849 for his refractor, to which he added a transit instrument.

Pritchard was full of original ability. He minimized rote learning, and taught mathematics and science (including botany and chemistry) in a practical way, zealous to inculcate the habit of thinking.[39] John Herschel, George Airy, and Charles Darwin all sent their sons to his school, and in 1866 nearly a hundred former pupils attended a banquet for him.[40] In Clapham, Pritchard had a coterie of scientific neighbours who shared a strong link between astronomy, microscopy, and religion. Sir George Grove, the distinguished musical encyclopaedist and a close friend of Pritchard's, had been a pupil of his at both Stockwell and Clapham, and recalled:[41]

> I had been at two other schools which were considered good of their class; but with Pritchard the atmosphere was very different. The master was younger and more sympathetic, and full of a wider knowledge than I had before dreamed of; also, he had a great power of explanation and illustration, and took constant interest in his boys. One or two things I had not met before, and they made a great impression on me. First, the mathematics and the natural philosophy. These were taught in a practical and interesting way, and as connected with common life – not with the abstract world – which made them always fresh. Some of his precepts and examples recur to me almost daily.

A man of faith who had been ordained in 1834, Charles Pritchard never held a curacy. He preached so many times for the British Association that he was known as its 'Chaplain', and was widely respected as a scientist preacher who specialized in reconciling nature with Revelation, and who stood staunchly for the established church against the Tractarians. Many scientific colleagues found this reconciliation intellectually challenging, and Pritchard corresponded frankly with Airy, Darwin, and Adams, among others. Always taking great care over his sermons to the British Association, he delivered the prestigious Hulsean theological lectures in Cambridge in 1867, and also preached in Westminster Abbey, in Salisbury, Hereford, and Worcester Cathedrals, and later in a series at St Mary's, Oxford's University Church.

Pritchard joined the Royal Astronomical Society in 1849, was elected to the Council in 1856, and served continuously until 1877 and again from 1883 to 1887. In 1860–61 he was vice-President to Robert Main's presidency, and was then secretary from 1862 to 1866 and president from 1866 to 1868. A colleague and friend there was Warren De la Rue, 'the father of celestial photography in England', and in 1860 Pritchard was a member of an eclipse expedition to Spain organized by De la Rue. Pritchard is glimpsed among the chief personalities during the 1860s:[42]

> Airy, a dominating figure, unbending and gruff; Adams, clear-minded, quiet and helpful; Pritchard, lively and sympathetic; Lee stately and courteous; Carrington, impetuous; Huggins, careful and judicial; and De la Rue, a man of order and energy, on cordial terms with everyone.

The Savilian election of 1869

When William Donkin died in November 1869, Henry Savile's unchanged statutes faced his electors with a challenge. The University had offered no observatory infrastructure to attract an experienced astronomer, nor the academic need for the advanced teaching that would make a professor welcomed among his tutorial colleagues. The council of the Royal Astronomical Society regarded Oxford's Savilian chair as merely a teaching post, and so Airy did not concern himself with it.

Meanwhile, the Devonshire Commission was taking evidence. None of the 'most distinguished' electors was a scientist, nor (except for the Warden of New College) was resident in the University, and so the new Chancellor, the Marquis of Salisbury, felt obliged to seek the advice of Francis Leighton, the Vice-Chancellor since 1866. Leighton knew that in future the Radcliffe Observer would be an elector for the Sedleian chair, and this may have prompted him to seek Main's advice.

It is not known who encouraged Charles Pritchard to apply for the Savilian chair in 1869, but circumstantial evidence suggests that Robert Main may have been instrumental. Pritchard was then aged 61 and had no thought of seeking employment. Also, the Astronomical Society's president, Admiral Russell H. Manners, who had worked with Pritchard and knew him well, had retired. But apparently Pritchard was urged to apply by his friends in the Society, while his future assistant, William Plummer, recalled that Pritchard had suggested that his sermons to the British Association had 'brought him to the notice, and subsequently to the friendship, of Lord Hatherley', who as Lord High Chancellor was an elector to the Savilian chair, and that it was due to his good offices that he was chosen.[43]

Pritchard was duly elected in February 1870, and was immediately invited to matriculate from New College, from where he then proceeded to his MA degree by decree in March. New College completed his welcome by finally paying £150 towards his stipend. He moved his young family to Oxford in the autumn, renting the house at 8 Keble Road.

After taking the chair in 1870, Pritchard promptly dedicated considerable effort to working with his friend Robert Main to co-edit John Herschel's *Catalogue of 10,300 Multiple and Double Stars*, published by the Royal Astronomical Society in 1874. Main and Pritchard were good friends, and Pritchard was later a mourner at Main's funeral in 1878 and wrote his obituary for the Society.

From his appointment in 1870, Charles Pritchard brought an energy to the Savilian chair of astronomy that had not been seen for many years. As well as being a superb teacher, he detailed and supervised the building of the new University Observatory in the University Parks.

A new observatory

After his appointment, Charles Pritchard repaired the Museum Observatory and Donkin's old 5-foot transit and clock and small altazimuth, and in 1871 he began to offer instruction, but the Observatory was too small to accommodate his own portable equatorial. As Turner recalled:[44]

> The Professor's work soon resolved itself into two parts – the delivery of lectures on Mathematical Astronomy (Newton's *Principia*, Lunar and Planetary Theories, with the purely mathematical parts of spherical astronomy), and the instruction in practical astronomy (which included calculating orbits, transits, parallax, and longitude).

Pritchard found the Observatory useless, but a close colleague who had 'outside the limelight ... steered Oxford science through the changes of the 1850s and 1860s' was the professor of geology, John Phillips, who had been Keeper of the University Museum since 1860 and was a keen astronomer.[45] Between 1868 and 1870 Phillips had steered the debate that persuaded Congregation to accept the physics laboratory, and he now had no difficulty in proposing honorary degrees for Pritchard's Astronomical Society friends, De la Rue, John Gassiot, and William Huggins.

In June 1872 Phillips offered to donate to the University his own 6-inch Cook refractor, provided that it was used and that the Royal Society (which had funded his observatory) agreed. Phillips's intention was that it would enable Pritchard to supervise some long-term research,

Warren De la Rue, co-founder of the Observatory of 1875.

so that astronomy might become institutionalized; this proposal was exciting for Pritchard, whose efforts had focused on teaching. Discussion between Pritchard and Phillips developed into testing whether the University might provide a permanent facility. Pritchard later wrote that 'it was very soon represented to me that if I could but have a little patience I should get all that I desired'.[46]

Phillips recommended the course of action – to find the largest refractor available for early delivery and its cost; the size of the instrument would then determine the size of the dome, height of the tower, its location, and estimated cost. Howard Grubb of Dublin had a 12.25-inch object glass available, and so a 'Plan of the Museum Site', dated January 1873, was drawn up by Phillips so as not to encroach on the University Parks. In early February Pritchard sent the plan and estimates to Convocation, saying that without it,[47]

> Astronomy can not be taught in a manner worthy of the present state of scientific education in Oxford, or of the ancient fame we inherit from Halley and Bradley.

On 4 March Convocation unanimously voted £2500. The Observatory would be a symbol of progress, and easier to digest than reforms of fellowships and scholarships. The astronomical world was astonished by the news. Shortly afterwards, Pritchard wrote:[48]

> Dining with Mr De la Rue . . . I was detailing the great liberality of the University when my friend stopped me and said he thought it would be a fitting thing for his great telescope to go over to Oxford.

The New Savilian Observatory for Astronomical Physics, 1875.

In June De la Rue offered his 13-inch photographic reflector and the contents of his observatory, on condition that it was properly housed and used. By 17 November 1873 the University had formally accepted the gift, committing £1000 for the second tower, £500 for a connecting corridor, and £200 a year for running costs and the stipend of an assistant. To emphasize the shift in character of the new institution, Pritchard renamed it 'The New Savilian Observatory for Astronomical Physics at Oxford'.

Charles Pritchard's arrival at Oxford had been made potent by government enquiry, his ideals stimulated by Phillips's proposed gift and expertise when negotiating the University's internal networks. Thereafter Pritchard's aspirations and management skills, sustained by De la Rue's continued generosity, ran their course.

In February 1873 Pritchard had promised to found a school of astronomical physics with an instrument suitable for teaching, but also with all accessories to enable research. In practice, De la Rue's gift of the photographic instrument (and in particular the £100 salary of an assistant for four years to operate it) transformed the whole plan. Crucially, for the long term and apart from doubling the size of the Observatory and its amenities, the University made the major concession of granting a new location that was further into the Parks, away from the major restriction of the high sightline imposed by the Museum upon the original plan for building beside it on the very limited Museum site. In a letter to the *Times* on 3 January 1874, Pritchard announced that 'the University has led in England for the first foundation of an observatory for astronomical physics'.

The Observatory was completed in November 1875 for £5090 14s. 8d, at £1000 over budget, while from January 1875 the University's annual grant to the new observatory was only £300. William Plummer had joined in September 1874 with a stipend of £160, and John Mullis joined as a 'servant' handyman at £50 a year and proved invaluable. In 1874 Gassiot donated a splendid solar spectroscope by John Browning. De la Rue's subsidy then brought Charles Jenkins from the Royal Observatory as Second Assistant.

Even without curriculum reform, the Observatory indicated a significant return towards research-led teaching, because its direction was formally a responsibility of the Savilian professor, and for the first time, the University had made a long-term commitment to astronomy. A new statute in 1875 separated the Observatory from the Museum delegacy, giving it institutional independence and crucially placing it under the superintendence of a Board of Visitors comprising the Vice-Chancellor, the proctors, the Astronomer Royal, the director of the Cambridge Observatory, and the Radcliffe Observer, with De la Rue and four members of Congregation. They met once a term, and each year the Savilian Professor reported to them, his report then being published in Oxford and by the Royal Astronomical Society. The professor was henceforth a man of science committed specifically to astronomy, not only a teacher, and by his presence in the Astronomical Society and the Royal Society he raised the University's profile there too.

Pritchard's lecturing

The Observatory was completed to great fanfare. Pritchard urged the University to give astronomy its own honours school within the Natural Sciences, and to provide three Studentships of £100 per year. Unsuccessful, he accepted that its reputation must be made in research. During his tenure he sent fifty astronomical papers to learned societies. He participated in the Ashmolean Society, was a founder member in November 1888 of the Oxford Mathematical Society, and also in 1890 of the British Astronomical Association which catered for amateurs unable to join the Royal Astronomical Society.

Pritchard never tired of teaching. For the academic year of 1881–82 he offered to students lunar and planetary theory and elementary principles; for graduates, tutors, and lay schoolteachers on three occasions he showed how geometry could be taught using examples from astronomy. For University members, or for the general public lacking mathematics but interested in astronomy, he advertised lectures on the mathematical, observational, and physical nature of comets, but not a single member of the University attended. From 1882 Pritchard did not mention lecturing in his annual reports, concentrating instead on research.[49] To 'bring astronomy into Oxford', Pritchard gave popular public lantern-slide lectures.

Unable to influence curricular reform, Pritchard had the satisfaction of seeing the 1877 Act reforming the statutes for the University and colleges forced through in 1882. The Savilian electors were revised to be the Chancellor, the president of the Royal Society, the Astronomer Royal, the Radcliffe Observer, the Warden of New College, and nominees from New College and the Hebdomadal Council, so at last there was scientific influence to safeguard the chair. The Savilian appointments were formally annexed to New College, whose statutes now created a new class of 'professor–fellows', in addition to the tutorial and other fellows. The professor would receive £400 from the Savilian fund, plus 'the emoluments of an ordinary fellow' which was fixed at £200 net.[50] Pritchard had at last attained his fellowship by right in 1883, with a regular stipend of £600.

Knowing the low ability of the University's students in mathematics, Pritchard did not press for curricular reform, and so engendered no conflict with college tutors. De la Rue's sustained generosity also saved him from making controversial demands for money. As Turner in 1896 summarised Pritchard's pragmatism:[51]

> The new observatory was prosperous ... in the direction of research ... [it was] a movement in the line of least resistance ... [although] the lavish expenditure of the

This caricature, entitled 'A Star in the East 1883', shows Charles Pritchard on an expedition to Egypt to investigate the absorption of light by the atmosphere. It was partly for his work on this topic that he was awarded the Gold Medal of the Royal Astronomical Society in 1886.

University upon scientific institutions and apparatus has not yet [in 1896] produced any very startling result in the schools.

Pritchard's research

Pritchard was the first British professor–director to adopt photography and astrophysics. He had little practical experience of research, but through his offices in the Royal Astronomical Society he had considerable knowledge of it. Between 1877 and 1881 he enabled Charles Jenkins to use the De la Rue reflector to investigate lunar libration, and measured the photographs with a superb micrometer donated by De la Rue, but chose not to publish the results. He then sought micrometer measurements of the photographed positions of forty stars in the Pleiades cluster to compare them with F. W. Bessel's visual observations made fifty years previously (see Chapter 4). To standardize the various estimates of the brightness of these stars, he adapted a new principle (using a wedge photometer designed by William R. Dawes) and defended his results. Pritchard and Plummer published papers on double stars, the Pleiades, and instrumental researches.

Following this trial, his next major research addressed the fundamental problem of determining the actual magnitude or brightness of all those stars visible to the naked eye in more than half the entire sky. From 1881 to 1885 Plummer made photometric measurements of 2784 stars from the pole to 10 degrees south of the equator, while Pritchard devoted two full years to publishing their catalogue, the *Uranometria nova Oxoniensis*.

In 1885 Pritchard was elected to the Council of the Royal Society and was appointed to its Solar Physics committee. In 1886 he was awarded the Royal Astronomical Society's Gold Medal for his photometry, jointly with Edward Pickering of the Harvard Observatory who had started earlier with a different method. Pritchard was also delighted to be elected to an honorary fellowship at St John's College, Cambridge, in 1886.

Charles Pritchard was a pioneer in proving the method of determining a star's distance by *photographic* parallax; this was innovative, and was his most important work. He had wondered about applying photography to that detection which had eluded Bradley, and in 1886 from two hundred plates he derived a parallax for the star 61 Cygni, and then from 1886 to 1889 he studied an additional twenty-eight brighter stars to the second magnitude. Pritchard was persuading astronomers to take accurate positional measurements from photographic plates, and in 1892 this parallax work earned him the Royal Society's Royal Medal. All of these observations had been made by Plummer and Jenkins, and he gave them full credit.

In 1886 he gave the Royal Society a lengthy description of his researches into stellar photography. He was one of the first to prove the method and precision measurement from the dry-plate photography proposed for the Astrographic Catalogue and the *Carte du Ciel* plan for eighteen observatories to photograph and catalogue the whole sky, the first large international co-operative programme in astronomy (see Chapter 7). Turner again referred to De la Rue's co-founder role, and to the joy with which Pritchard must have received the following letter from De la Rue, dated 16 June 1887:

> My Dear Pritchard,
>
> It gave me very great pleasure to learn that the Visitors of the Oxford University Observatory, recognising the value of your photo-astronomical work, are about to request the University to vote for five years the sum of £600 per annum to enable you to continue the work. I am myself deeply sensible of the value of your work, and, provided the University will accept my offer, I will present to it, in your honour, the sum of £500, to enable it to procure the conversion of the Grubb telescope into an astronomical photographic of the very highest order.

De la Rue's offer was accepted by the University's Convocation on 1 November 1887. On the following day he sent a cheque to the Vice-Chancellor, requesting an inscription to be affixed to the telescope, saying 'The gift of Warren De La Rue in honour of Professor Pritchard'. De la Rue died two years later on 19 April 1889, at the age of 74, and as Turner observed:[52]

> It would be difficult to over-estimate the value of his help, so generous and so well timed. On at least two occasions his liberality made all the difference. In 1874 it converted a refracting telescope into a well-equipped observatory; and in 1887 it enabled that observatory to retain its place among those engaged in the very newest and most important work, when it otherwise might have had to fall behind.

Pritchard promptly committed the Observatory to the *Carte du Ciel* project. He then played a vital leading role in pressing Howard Grubb to produce absolutely identical object glasses for the astrographs. He had developed real expertise, and was respected for it by William Christie, the Astronomer Royal, and his assistant Herbert Turner at the Royal Observatory in Greenwich.

Pritchard had achieved the creation of a large observatory as a separate institution funded and owned by the University, accountable to it via a Board of Visitors, and thereby ensuring

Oxford University's 12¼-inch Grubb refractor of 1875.

both public and professional awareness. He had put Oxford University back on the world map of astronomy – but he still had no graduate students and no laboratory. During his seventeen years the only augmentation of instrumental power was De la Rue's astrographic camera mounted coaxially upon the Grubb refractor.

William Plummer, the lead observer and effective deputy director, had been a zealous and highly competent observer and astronomer, but he left Oxford in 1892 to take direction of the Bidston Observatory in Liverpool. Indispensable to Pritchard and to the success of the Observatory, he was recognized by an Oxford honorary MA degree in 1892. Plummer had been assisted by Charles Jenkins in making all the observations and measurements for the four volumes of Oxford observations that Pritchard directed and analysed, discussed, and published. A loyal and invaluable astronomer who had made a great contribution,[53] Jenkins left the Observatory after Pritchard's death in 1893.

In Pritchard's last years his eyesight was failing, mobility was a problem, and he was cared for by his daughter Ada. From 1889 he was conveyed to the Observatory in his bath-chair and gave his prescribed lectures. In 1891 he suddenly lost his wife Rosalind, and sought distraction in work. Indomitable, he lectured during that October term, and again in 1892. His thoughts turned again to further photographic investigation of the Pleiades. During these years Henry

Acland, 'kind, genial, broad-minded, curious' was a frequent visitor, as was Professor Sylvester, 'warmest and most loyal'.[54] Pritchard died at his home in Oxford on 28 May 1893, aged 85. As Turner recalled:[55]

> The Oxford University Observatory having been built, and made thoroughly efficient; the *Uranometria Nova Oxoniensis* having been published; the parallaxes of the second magnitude stars having been determined; and the Oxford share of the astrographic chart having been successfully incepted – at this point the master hand was withdrawn from the work.

Conclusion

The tenures of the three Victorian professors span a half-century, at a time when Oxford University had to be compelled by government to any reform, and implementation was often resisted and delayed. Did each incumbent seek and manage to make a mark in advancing mathematical teaching and institutional development, or contribute to astronomy that would enhance the University's reputation and attract investment and students?

All three professors were appointed by the government–University's Anglican elite, not one of whom was a man of science. But context is everything, and Rigaud's death in 1839 focused the University and electors on requiring an impeccably qualified Anglican conservative mathematician, while the Radcliffe Trustees had an uncompromisable need for a first-class astronomer. George Johnson was the University's best fit. For Robert Peel he had lacked the basic qualification, but in hurriedly appointing him the University gave no thought to the interests of their great benefactor. Bereft by the schism, Johnson made no attempt to contribute to astronomy in Oxford.

By 1842 the University's priority was still the need for a reliable Anglican. William Donkin was the outstanding scholar with some knowledge of astronomy, although it was not an area of research that he had intended to pursue. He did his duty, won friends, and gained influence. But Donkin had neither the extrovert personality for campaigning, nor the health to battle the tide of Oxford indifference. He advanced mathematics within his University which had no need of astronomy, but with the University paying for the Museum Observatory and then rejecting the recombination of posts, two determining precedents were set.

In 1870 Pritchard was delighted to be appointed to the chair of Wren, Bradley, and Hornsby. The first incumbent since Hornsby to tackle his dual duties with zeal and enthusiasm, he found an immediate ally in John Phillips who, capitalizing on the Devonshire Commission's scrutiny, was the catalyst for the new observatory. De la Rue was its co-founder and his decisive interventions transformed the project. Thirty-four years after the schism with the Trustees, the University had at last made its first serious cash investment in astronomy in Oxford.

The three Victorian professors reflected the social complexion of British astronomy: a cleric (Johnson) unsuited to the responsibility, the academic professional (Donkin) who saw himself as an undergraduate teacher, and a geological amateur astronomer, paper manufacturer, and former schoolmaster (Pritchard), who began a *real* research institution in Oxford.[56]

In reality, the new observatory was a hybrid because its photographic telescope of 1849 was near obsolescence. It also had no laboratory and was insufficiently funded for astrophysical work, but had no graduate students nor scholarships to attract them. Nevertheless, it had hugely important long-term consequences for astronomy in Oxford. Pritchard embraced the new opportunities of photography applied to stellar astronomy, with a staff devoted to him who made it possible to achieve two prestigious awards. These put Oxford's astronomy back on the international map, so that in 1893 the chair was sufficiently attractive to a first-class astronomer, Herbert Hall Turner, who was dedicated to research and built international networks.

In the long view, the 1839 débacle had exposed the issue of Observatory ownership. The completion of the University Observatory in 1875 created a proximate pair of observatories within just half a mile of each other, and this made it more difficult for the professors to gain instrument renewal. Pritchard's remarkably good use of the Observatory restored the principle of research-led teaching, but weak mathematics before 1928 disadvantaged Oxford.

Whenever it became possible to combine spectroscopy with photography, the new specialities in astrophysics needed graduates who were trained in physics and chemistry. Ageing Grand Amateurs sought for where they could best donate their powerful instruments.[55] London University's Mill Hill Observatory of 1924 gladly accepted the Radcliffe's 24"/18" photo-visual double refractor of 1906, after the Trustees in 1935 had first offered it to Oxford which could not afford to accept and mount it.

However, being free of University obligations the Radcliffe Observers developed their own aspirations, which led between 1935 and 1948 to their commissioning a powerful 74-inch Grubb reflector on a fine-weather site in South Africa. The Trustees then offered research travelling fellowships to Oxford graduates, found first-class candidates there, and the century-long schism was thereby healed.

Herbert Hall Turner with the Oxford astrographic telescope.

CHAPTER 7

The 20th century

LEE T. MACDONALD

Just three individuals held the Savilian Professorship of Astronomy between 1894 and 1988: Herbert Turner (from 1894 to 1930), Harry Plaskett (from 1932 to 1960), and Donald Blackwell (from 1960 to 1988). Yet during this same period, astronomy changed beyond recognition. In 1894, as it had been for centuries, astronomy at professional observatories in the British Isles was predominantly the science of positional measurement. But during the tenure of these three Savilian Professors, it became centred on astrophysics – the study of the properties of the celestial bodies themselves, especially with the use of spectroscopy and photography, the latter being replaced by electronic media towards the end of the period. Also, by 1988, the raw observational data of astrophysics was generally being acquired far from Oxford, at clear-sky mountain-top observatories overseas, or by telescopes and specialized satellites in space. This chapter describes how Turner, Plaskett, and Blackwell contributed to this changing world of astronomy.

Herbert Turner

Herbert Hall Turner was born in Leeds on 13 August 1861, the son of an artist. He won a scholarship to Clifton College, a major public school near Bristol, where between 1874 and 1879 he received a wide education in mathematics and the physical sciences in addition to the classical subjects so often associated with public schools. In 1879 he won a scholarship to Trinity College, Cambridge, to read for that university's famous Mathematical Tripos. By the mid-19th century this Tripos had become established as the essential training ground for anyone in Britain considering a career in astronomy, especially at the Royal Observatory in Greenwich. Many of the problems set in Tripos examination papers were in mathematical astronomy and related subjects,[1] but at this stage Turner had expressed no particular interest in observational astronomy.

In 1882 Turner graduated as 'Second Wrangler', the candidate in the Mathematical Tripos with the second-highest marks in his year. Although he had received a privileged education, he was not from a wealthy background and needed scholarships in order to pursue his studies, both before and after graduation. In his graduation year he won the University of Cambridge's Sheepshanks exhibition for astronomy, which provided a stipend of £30 – £40 per year and came with membership of Trinity College.[2]

Lee T. Macdonald, *The 20th century*. In: *Oxford's Savilian Professors of Astronomy*. Edited by: Robin Wilson and Steven Balbus, Oxford University Press. © Oxford University Press (2025). DOI: 10.1093/oso/9780198894292.003.0007

This launched Turner's career in astronomy – in particular, bringing him to the attention of William Christie, Astronomer Royal since 1881; indeed, it seems to have been Christie himself who brought Turner into the astronomical fold. Soon after George Airy became Astronomer Royal in 1835, he had instigated a policy of appointing a recent top Cambridge Wrangler as his deputy at Greenwich, known as his 'Chief Assistant'. Airy retired in 1881 and Christie, his final Chief Assistant, succeeded him. Christie then simply promoted Edwin Dunkin, one of the assistant astronomers at Greenwich, to be his Chief Assistant, in order to keep the Royal Observatory running and to ensure a smooth transition between the Astronomers Royal. However, Dunkin was due to retire in 1884, and in the run-up to his departure, Christie resumed Airy's policy of appointing a Cambridge Wrangler; indeed, according to Turner's later obituary of Christie, Christie volunteered to be an examiner for the Sheepshanks exhibition in order to look for a potential Chief Assistant among its candidates.[3] He chose Turner, who started working at Greenwich on 25 August 1884.[4]

As Christie's deputy, Turner was responsible for managing the Royal Observatory and its staff in his absence. His own research interests initially focused on various aspects of the positional astronomy that formed the traditional 'bread-and-butter' of the observatory's work, including observational problems using the Airy transit circle – a telescope used for measuring the positions of stars as they crossed the meridian, and which formed the basis of Britain's time service – and the 'personal equation' (the response time of an individual observer to the passage of a star behind the wires in the transit circle eyepiece). Turner also became an enthusiastic observer of solar eclipses and, along with E. Walter Maunder, head of the Royal Observatory's solar department, he visited Grenada in the West Indies to observe the total solar eclipse in August 1886, making straightforward visual observations of the spectrum of the Sun's corona. In the following year, he experienced an all-too-common disappointment encountered by eclipse observers, when he travelled to Russia to see an eclipse that was covered by clouds.[5]

The *Carte du Ciel*

The area of astronomy in which Turner made his greatest mark while at Greenwich was stellar photography, and specifically a project that became known as the *Carte du Ciel*, or 'Map of the Heavens'. The late 1870s had seen the invention of the 'dry' photographic plate that made it practical to photograph the night sky with the long exposures needed to record faint stars. The first observer to see the dry plate's potential for mapping the stars was David Gill, Her Majesty's Astronomer at the Cape of Good Hope since 1879, and in 1882 he used dry plates with a portrait lens mounted on a clock-driven telescope to take some wide-angle photographs of that year's bright comet. In addition to the comet's great tail, Gill recorded huge numbers of faint stars, an achievement that inspired him to make a photographic survey of the southern sky.

At around the same time, the brothers Paul and Prosper Henry at the Paris Observatory, frustrated by the slowness of visual methods for mapping stars, took some remarkably successful photographs of stars with a 13-inch refractor designed for photography. This led to the Paris Observatory's director, Ernest Mouchez, corresponding with Gill about a possible survey of the whole sky by a number of participating observatories. In April 1887 fifty-six astronomers from across the globe met at a congress in Paris and agreed to begin such a collaborative survey (the *Carte du Ciel*), with each participating observatory photographing an allocated zone of the sky, using identical 13-inch photographic refractors. One of the astronomers taking part in this conference was Christie, who agreed that the Royal Observatory should photograph the zone between +65° declination and the north celestial pole. He ordered a 13-inch photographic

Participants in the Paris International Astrographic Congress of 1887.

refractor from Howard Grubb, the Dublin telescope maker, as did Oxford's Savilian professor Charles Pritchard, who agreed to photograph the sky between 25° and 31° north declination.[6]

Although photography might seem to simplify mapping the stars, compiling a usable star catalogue from photographs presented the observatory staff with a great deal of work. They had to measure the positions of potentially thousands of star images on each plate, and then reduce these measured positions to right ascension and declination coordinates that astronomers could use. Turner made critical contributions to both of these problems. He invented a new type of micrometer that speeded up the measurement of star positions on plates, and by late 1893 he was also developing a system of coordinates for the measured star positions that could easily be converted to right ascension and declination with a minimum of calculation.[7]

In 1893 Charles Pritchard, Oxford's Savilian professor of astronomy, died while still in office, and Turner, only 32 years old but a rapidly rising figure in British astronomy, applied for the vacant post. As with his election to the Sheepshanks exhibition and the Chief Assistantship at Greenwich, Turner was in a strong position, in that Christie was one of the seven electors who met on 1 December 1893 at the London home of Lord Salisbury, Oxford University's Chancellor, to elect the next Savilian professor. In addition to Christie and Lord Salisbury, the other electors included the renowned physicist Lord Kelvin and Edward Stone, director of Oxford's Radcliffe Observatory, which by now was managed independently of the University by the Radcliffe Trustees. The meeting unanimously elected Turner from a pool of six candidates.[8]

Turner's expertise in the *Carte du Ciel* surely did much to win him the favour of the electors. When Pritchard had agreed that the Oxford University Observatory should take part in the project, Warren De la Rue, an astronomical photography pioneer (see Chapter 6), had paid for the requisite 13-inch photographic refractor and had arranged for it to be mounted in a dome at the University Observatory alongside an existing 12-inch visual refractor. By the time of Pritchard's death, Oxford was firmly committed to taking part in the *Carte du Ciel*.[9] The importance of Turner's experience in this is strongly suggested by his printed application to the electors. In the final paragraph of his application, after listing his scientific papers to date, Turner described his work on the *Carte du Ciel* at Greenwich, noting that this was also 'the chosen work of the Oxford University Observatory at the present time', and expressing the hope that, if elected, 'this experience may enable me to take up the work of the Observatory with the least possible delay or discontinuity'. He also sent the electors an offprint of a new paper that he had written about his method for reducing the coordinates of star positions, with a note to the effect that it 'has special reference to the last paragraph of my application'.[10]

Turner's tenure at Oxford started in February 1894. As Savilian professor of astronomy Turner was a Fellow of New College, and initially lived in college rooms until his marriage in 1899 to Agnes Whyte; the couple had one daughter.[11] In addition to his statutory teaching duties, Turner's first research task was to complete the work on Oxford's *Carte du Ciel* zone, begun under Pritchard.

The *Carte du Ciel* project was in two parts. The whole of an observatory's allocated zone was to be photographed using short exposures, with the purpose of extracting the positions and magnitudes of stars down to magnitude 11, for publication in the *Astrographic Catalogue*. A second set of photographs was then to be taken with much longer exposures, to produce an atlas showing stars to magnitude 14. Known as the *Astrographic Chart*, this was to take the form of photographic prints of the sky in each observatory's zone, to be copied and distributed to fifty other observatories. Covering Oxford's zone would involve taking two sets of 1180 plates, one for the *Astrographic Catalogue* and the other for the *Astrographic Chart*.[12]

The Royal Observatory at Greenwich was ideally suited to such a labour-intensive project. It had a large number of assistant astronomers and an especially large staff of temporary 'computers' – mostly teenage boys, hired to perform elementary but tedious calculations, as well as clerical work. Such a large staff was sufficient for taking Greenwich's share of the plates, and then measuring and reducing the star positions. Moreover, by the late 19th century Greenwich had long been established as a 'factory' that produced and published quality astronomical data in a timely manner.[13]

But at Oxford, Turner had no such resources. Established in the 1870s, Oxford's University Observatory had a tiny staff and budget. Warren De la Rue, who paid for the 13-inch astrographic refractor, had died in 1889, bringing to an end this important private source of income, and Turner had to rely on a tiny annual sum from Oxford University and small grants from the Royal Society. Like several other participants in the project, Turner did not even attempt the long exposures, concentrating instead on the short exposures for the *Astrographic Catalogue*. In early 1901 he was complaining to Christie that in order to finish the measurements and reductions for Oxford's *Astrographic Catalogue* zone, he would need to apply for a further £300 from the Royal Society Government Grant on top of the £750 that he had received five years earlier.[14]

Turner was never much of an observer, as such. He preferred instead to be the scientific brain behind a project, solving the technical problems – such as that of reducing the stellar coordinates – and generally directing the University Observatory. Most of the plates for Oxford's contribution to the *Carte du Ciel* were taken by Frank Bellamy, who in 1892 had been appointed 'First Assistant', the most senior astronomer at the Observatory apart from the Savilian professor, and would remain loyally in that position throughout Turner's tenure of the chair. The Observatory also had a succession of people working as 'Second Assistant', initially funded by De la Rue and then by increasingly precarious grants from other sources.

As at Greenwich, much of the hard work in measuring and reducing the star positions was done by computers. Prominent among these was Ethel Bellamy, Frank Bellamy's niece. Chronically short of funds, Turner frequently outsourced the measuring work to volunteers, among them the Yorkshire gardener and amateur astronomer Tom John Moore, who measured the star positions on 301 of the Oxford zone's 1180 plates.[15] For a while, Turner took advantage of his College membership to employ choirboys from New College Chapel to help with the measurements. In this way, Turner and his helpers finished the short-exposure plates in 1904, and Oxford's part of the *Astrographic Catalogue* was completed and published in 1911.[16]

Frank Bellamy, the First Assistant at the University Observatory under Turner, with the 13-inch De la Rue reflector.

Other interests

Turner's astronomical interests ranged widely. He undertook analyses of his *Carte du Ciel* plates with a view to tackling the 'sidereal problem' – determining the distances, motions, and distribution of the stars through what, until the discovery of external galaxies in the 1920s, astronomers still generally called 'the Universe'. The sidereal problem was a cutting-edge research topic for many astronomers in the early 20th century, and Turner was at the forefront of developments in it, being keen on using data for detailed analysis, rather than simply accumulating observations for their own sake. According to an obituary by Ralph Sampson, the Astronomer Royal for Scotland, Turner considered that no one should amass observations for more than two years without printing them, or for more than five years without discussing them.[17]

In addition to his work based on *Carte du Ciel* data, he statistically analysed observations of variable stars in order to look for periodicities. He remained an enthusiastic observer of solar eclipses after his move to Oxford, and took part in expeditions to observe eclipses in Japan (1896), India (1898), Algeria (1900), and Egypt (1905). He also observed, from Southport

(Lancashire) the eclipse of June 1927, the only total solar eclipse to be visible from mainland Britain in his lifetime.

In the 1890s Turner was an early advocate of the *coelostat* – a mirror, mounted on a small clock-driven equatorial mount, that reflected the Sun's light into a horizontally mounted telescope, thereby removing the need to transport heavy telescope mounts to remote eclipse sites.[18] Although he never had the resources for work in the emerging science of astrophysics, he was very interested in it, and especially in solar physics. In 1904 he attended the first meeting of the International Union for Cooperation in Solar Research (IUCSR), held in St Louis, Missouri, and then hosted that organization's second meeting at New College, Oxford, in the following year.[19]

Turner carried out all his work against a backdrop of severe financial constraints imposed by Oxford University. This was symbolized by his attempts, in 1900–01 and again in 1907, to persuade the University to build a house next to the University Observatory, believing that an on-site residence would enable the professor to open the observatory very quickly whenever Oxford's unpredictable weather cleared, but despite some determined and well-publicized campaigning, this house was never built. The Observatory was located on the edge of the University Parks, whose wardens were firmly opposed to any further building work. In addition, the University itself was short of funds, and some of its senior members resented spending good money on the University Observatory at a time when the Radcliffe Observatory, recently given a new lease of life with a large photographic refractor, was doing good work less than a mile away.[20]

Roger Hutchins has suggested that Turner's failure to persuade Oxford University to build a house next to the observatory is illustrative of an aspect of his personality: tact was not his strong point, and he could be irascible.[21] His best-known outburst occurred during the First World War. From 1888 to 1897 he had edited *The Observatory*, a journal originally founded by Christie in 1877, containing astronomical news as well as research papers. From 1894, soon after his move to Oxford, he wrote a regular column of astronomical news and gossip, titled 'From an Oxford Note-Book', and although the column was anonymous it was an open secret that its author was Turner. In the May 1916 issue of *The Observatory*, with the First World War at its height and stories of German atrocities circulating widely, Turner queried:

> But shall we be able to resume the use of German organizations in the near future? . . . Many of us do not see how, after such an exhibition, we can face the mockery of new understandings and undertakings with such a nation.

This triggered an angry exchange with Arthur Stanley Eddington – like Turner, a former Chief Assistant at Greenwich and now the holder of a prestigious academic chair, the Plumian professorship of astronomy at Cambridge. Eddington, a Quaker pacifist, responded that it would be wrong to cut off communication with German colleagues, believing that it was not simply German atrocities, but militarism and aggression on both sides, that had caused the war, and claiming that 'The worship of force, love of empire, a narrow patriotism, and the perversion of science have brought the world to disaster'. Turner's response was uncompromising:[22]

> My reply is that the *facts* stand in the way – hard, horrible facts, such as we should not have believed possible before the war.

These examples of Turner's irascibility detract from many other aspects of his character. Turner was normally very internationalist in his astronomical outlook, as is evidenced

Herbert Hall Turner (1861–1930).

by his work in the *Carte du Ciel* and the International Union for Cooperation in Solar Research. In addition to his own contribution to the *Carte du Ciel*, he supervised the reductions of the star measurements from its plates taken at the Hyderabad Observatory in India and the Vatican Observatory, and after the First World War, he headed the newly founded International Astronomical Union's commission that took over responsibility for the *Carte du Ciel*.

Turner was also noted for his sociability. He was a *bon viveur* in the Royal Astronomical Society Club, a dining club whose members dined at the Athenaeum club after the monthly Royal Astronomical Society meetings. He served as the Society's secretary from 1892 to 1899, its president from 1903 to 1905, and its foreign secretary from 1919 until his death in 1930. At New College, he regularly played in an annual cricket match between the Senior Common Room and the choirboys. A dedicated member of the College's governing body, he assisted the Warden with inspecting farms owned by the College. His column 'From an Oxford Note-Book' was noted more for its humorous anecdotes than for wartime political rants.

Turner was also a keen popularizer of astronomy, spreading it far beyond academia. He gave many lectures to Workers' Education Association classes across Britain, using the long train journeys as an opportunity to work through proofs of his publications.[23] He also wrote several popular books on astronomy. One of these, *The Great Star Map* of 1912, remains the best contemporary account of the *Carte du Ciel* project in English for the non-specialist reader.[24]

Seismology

In the 21st century Herbert Turner is now perhaps best remembered for his work in a distinctly different field. Among his wide circle of friends was the geologist John Milne, one of a number of British men of science who were encouraged by the Japanese government to teach Western science in Japan in the late 19th century. Indeed, Milne sent Turner some advice about travel in Japan before Turner arrived there for the 1896 solar eclipse. During his years in Japan, Milne had become interested in earthquakes and had emerged as an early expert in what came to be called 'seismology'. After his return to Britain in 1895 (with a Japanese wife), Milne set up a private centre for seismological investigations at Shide on the Isle of Wight, to which observers across the British Empire began to send reports of earthquakes. Milne published summaries of these reports in his 'Shide circulars'.[25]

Turner soon became interested in the subject, serving with Milne on the committee for seismological investigation of the British Association for the Advancement of Science from 1898. In 1913 Milne died at the age of 62, and Turner agreed to take over the coordination of the seismological reports, seemingly from a gentlemanly loyalty to his old friend. To begin with, the seismological observations had been collated by volunteers working at Shide, but at the end of World War I Turner transferred the work to Oxford. Until 1928 these observations were collated and reports were compiled in the 'Students' Observatory' which was effectively a wooden shed with no heating; the work was then moved to an extension of the University Observatory. As with astronomy, Turner had no spare funds for this new venture. He received some financial support from James E. Crombie, a private benefactor, but otherwise he had to rely on what small grants he could secure from other sources in the difficult economic conditions of the time.[26]

Turner's interest in seismology may have been partly inspired by his instinct for analysing data and solving problems. As with his *Carte du Ciel* activities, where he was not content with merely collecting data, he started to collate the reports sent to him from the earthquake events themselves, rather than from the observing stations as had been done previously. He found innovative ways to calculate both the location of an earthquake's epicentre and the depth of its focus below the Earth's surface.

He also discovered 'deep-focus' earthquakes. Hitherto, seismologists had thought that earthquakes generally occurred close to the Earth's surface, but Turner demonstrated that some earthquakes had a focus nearly 400 kilometres below the surface. He became responsible for managing a seismological monitoring operation that by 1929 had received reports from 259 observing stations; these were now scattered all over the Earth, and not just the British Empire as had been the case in Milne's time.

In 1919 Turner became president of the seismology section of the International Union of Geodesy and Geophysics, in addition to his responsibilities as president of the International Astronomical Union's Astrographic Catalogue Commission (which was responsible for the *Carte du Ciel*). To calculate the epicentres and timings, Turner employed the mathematician Joseph Hughes as Seismology Assistant from 1923. Their results were published as the *International Seismological Summary*, compiled and distributed by Ethel Bellamy, and in 1934 Oxford University awarded her an honorary MA degree for her seismological and astronomical work.[27]

Turner's many other duties, when combined with the huge burden of the seismological reports and the persistent lack of funding, may well have led to his suffering from overwork.[28] He died suddenly of a cerebral haemorrhage on 20 August 1930, while he was chairing a meeting of the seismological section of the International Union of Geodesy and Geophysics in Stockholm.

By the time that Turner died, Oxford astronomy, and indeed the whole University, had become embroiled in an increasingly bitter controversy. In October 1923 Arthur Rambaut,

(*Left*) Herbert Turner with a seismology globe. (*Right*) Ethel Bellamy receiving her honorary MA degree in 1934.

the incumbent Radcliffe Observer, had died and was succeeded by Harold Knox-Shaw who had previously directed the Helwan Observatory in Egypt. Knox-Shaw was soon disappointed by the lack of usable clear nights in Oxford. So when, five years later, the motor-car magnate William Morris (Lord Nuffield) offered the Radcliffe Trustees £100,000 in compensation for removing the Radcliffe Observatory to make way for an extension to the neighbouring Radcliffe Infirmary, Knox-Shaw jumped at the chance. This large compensation payment would be sufficient to set up a new Radcliffe Observatory under the clear skies of South Africa.[29]

Turner had no objections to this idea – on the contrary, as early as 1907 he had suggested moving the Radcliffe Observatory to the Transvaal,[30] and he now began working with Knox-Shaw on a scheme for future cooperation between the two observatories. In April 1930 the Radcliffe Trustees decided to proceed with moving the observatory to South Africa, and on the same day Turner published a letter in *The Times* supporting the case for removal. But the plan aroused the ire of another strong character, Frederick Lindemann (later Lord Cherwell), professor of physics at Oxford since 1919, who objected to the Trustees' proposal on the grounds that it would involve exporting to South Africa money that ought to be spent on astronomy and related subjects within the University of Oxford.

However, Lindemann had his own agenda. He needed money to update the Clarendon physics laboratory, and he wanted to press the Radcliffe Observatory's large double refractor into service for his own research into the physics of the upper atmosphere. Lindemann

A memorial plaque in the New College cloisters commemorating the life of Herbert Turner.

responded to Turner's letter in *The Times* with a bitter reply, attacking the proposal to move the Radcliffe Observatory, and also Turner's own track record in astronomy and the allegedly obsolescent University Observatory. By midsummer 1930 Lindemann had persuaded the board of the Faculty of Physical Sciences to complain to the University's Hebdomadal Council that it would be 'deplorable' to divert the Radcliffe compensation money to South Africa.

Harry Plaskett

In the months following Turner's death in August 1930, Hebdomadal Council decided to postpone the election of Turner's successor for six months, while it waited for a committee to report on the future of astronomy in Oxford. In the meantime, partly through lobbying by Lindemann and by E. Arthur Milne, the newly appointed Rouse Ball professor of mathematics, the University's management was persuaded to challenge the Radcliffe Trustees in court.[31]

It was thus not until November 1931 that the University announced that the new Savilian Professor of Astronomy was to be the 38-year-old Canadian astrophysicist Harry Hemley Plaskett. Born in Toronto on 5 July 1893, he grew up in the world of astronomy. His father John Stanley Plaskett, who did much to make Canada one of the world's great astronomical nations, was the first director and intellectual leader of the Dominion Astrophysical Observatory at Victoria, British Columbia. Its 72-inch reflector was the second-largest telescope in the world when completed in 1918. Both father and son would later give Oxford's prestigious Halley lecture on astronomy in Oxford, in 1935 and 1965.[32]

Harry Plaskett attended state schools in Toronto and in Ottawa, where his father moved in 1903 to help with instrumentation for the new Dominion Observatory being built there. Although Harry was brought up in an astronomical household, it was only after enrolling at the University of Toronto in 1912 to read mathematics and physics that he began to focus on astronomy as a career. This turn owed much to his father, who let him observe with the Dominion Observatory solar telescope, and he used it for the research that he presented in his first published paper, on the Sun's rotation. By the time he completed his BA degree in 1916, he had become fascinated by spectroscopy, which would dominate his research career.

Harry Hemley Plaskett (1893–1980).

After finishing his degree, Plaskett served in France as a gunner in the Canadian Field Artillery until the end of World War I. He kept up his astronomical connections during the war years, staying at the Royal Observatory in Greenwich with the family of Frank Dyson, who had succeeded Christie as Astronomer Royal in 1910. After the Armistice, he spent some months working at Imperial College, London, with Alfred Fowler, the distinguished pioneer of astronomical spectroscopy. He then, through his father's influence, obtained a job at the new Dominion Astrophysical Observatory in British Columbia, where he made innovative use of the recently invented 'wedge' spectrophotometer to study a class of hot stars known as 'O-type' stars. He also brought himself up to speed on the latest mathematical and physical theory behind stellar astrophysics, a field that had recently been instigated in its modern guise by Arthur Eddington.

In 1921 Plaskett married Edith Alice Smith, with whom he had a daughter and a son, while his work brought him to the attention of Princeton University's Henry Russell, one of the leading theoreticians in stellar evolution. When Harlow Shapley, the recently appointed director of Harvard College Observatory, began looking for someone to lead his new 'graduate school' in astronomy, Russell recommended Plaskett.

Despite having no PhD degree, Harry Plaskett joined Harvard in 1928 as a lecturer, and by the time of his departure in 1932 he had become a full professor. During his years there, his research turned to solar spectroscopy, seeking to match his observations of Fraunhofer lines in the Sun's spectrum with current models of stellar atmospheres. In 1931 he published a major paper on the formation of magnesium *b* lines in the solar atmosphere, based on his spectroscopic observations.[33]

Harry Plaskett's appointment to Oxford owed much to the influence of the Rouse Ball professor Arthur Milne, who specialized in theoretical astrophysics, especially stellar structure and cosmology. Milne strongly believed that theorists and observers should work together and exchange ideas, and joined the chorus of those opposed to moving the Radcliffe Observatory, pressing instead for the establishment of an astrophysical institute in Oxford where such collaborative work could take place. In particular he wished the Oxford University Observatory to become a modern observatory that concentrated on astrophysics (meaning spectroscopic analysis), rather than on the traditional positional astronomy that had dominated the research agenda under Turner.[34]

Milne had first met Plaskett in 1929, when both attended a summer school in astronomy and physics at the University of Michigan, and had been hugely impressed by Plaskett's theoretical insight as well as his observational abilities. According to Milne's daughter and biographer, Meg Weston Smith, Milne influenced Plaskett's election 'at every stage', even though he was not on the board of electors. He persuaded Plaskett to apply for the position, despite the salary on offer being lower than what Plaskett was earning at Harvard. In October 1931 Milne wrote to the University's senior management, praising the 'provocative' quality of Plaskett's work, while Plaskett agreed to apply because he liked the idea of working with Milne. He arrived in Oxford to begin his new job in June 1932.[35]

Solar physics

In establishing his research agenda after starting at Oxford, Plaskett showed considerable political acumen. In November 1932 he presented the University Registrar with a proposal for a solar telescope to be erected in a revived and modernized University Observatory, regardless of what happened with the University's legal challenge to the Radcliffe Trustees. He decided to concentrate on solar physics with a relatively small instrument, partly because this was more realistic than observing faint objects with a large telescope in Oxford's indifferent climate. Moreover, at a projected cost of £10,000, his solar telescope would be far cheaper than the much larger instrument that would be needed for stellar spectroscopy. Presenting his proposal to the ever-impecunious University as the cheapest way of modernizing the observatory did much to convince the University to fund the solar telescope.[36]

Plaskett was genuinely very interested in solar physics, to which he had turned well before coming to Oxford. He believed that before studying the astrophysics of stars, it was essential to study the Earth's local star, the Sun, because it was the only star that could be examined in detail.[37] The solar telescope was completed in 1935 and was erected within the dome that had originally housed Warren De la Rue's 13-inch reflector.

Everyone in Oxford was now happy, except for Lindemann who never obtained the use of the Radcliffe refractor for upper-atmosphere physics. In July 1934 the University lost its legal case against the Trustees, who then proceeded to move the Radcliffe Observatory to South Africa. Eventually, after many delays, the Trustees built a 74-inch reflector at a site near Pretoria, which would long remain the largest telescope in the southern hemisphere. At the same time, the Radcliffe refractor was transferred to the University of London's new Mill Hill Observatory in north London, where it remains to this day.[38]

Although he had no particular interest in either, Plaskett inherited the two main lines of research that had been carried out under Turner – the *Carte du Ciel* and seismology. Meanwhile, Frank Bellamy had kept the observatory ticking over as acting director since Turner's death, and in 1932 when Plaskett took over, he was still working on the *Carte du Ciel* – now reducing position measurements of stars for the Potsdam Astrophysical Observatory's zone. Anxious to finish this task, Bellamy repeatedly clashed with Plaskett who was keen to steer the observatory towards solar physics. He resigned in January 1936 and died just two weeks later. Ethel Bellamy continued to compile reports for the *International Seismological Summary*, and the University paid her salary until she retired in 1947. By then, the British Association's Seismological Committee had agreed to transfer responsibility for their work from Oxford to Kew Observatory, where seismological observations had been carried out since 1898.[39]

From the mid-1930s, armed with his new solar telescope and with the politics of Oxford's observatories resolved to his satisfaction, Plaskett could at last concentrate on his solar research. While he was still at Harvard, he had begun studying solar 'granulation', the short-lived granule-like structures into which the Sun's gaseous surface appears to be broken. Using the new telescope at Oxford, Plaskett took further spectra of the granulation and used the results to demonstrate that the 'granules' are gaseous convection cells. He also obtained the best measurement yet of the temperature distribution at the base of the Sun's atmosphere. At the outbreak of World War II he was working on a mathematical model of the solar atmosphere consistent with observations. The plates on which the spectra were taken were all measured manually, and the positions (and therefore wavelengths) of absorption lines were measured using a hand-cranked moving microscope. All readings were recorded by hand, and calculations were performed using slide rules and, at least to begin with, a single hand-operated Brunsviga mechanical calculator.[40]

In addition to redirecting the research agenda at Oxford, Plaskett changed its research culture, recruiting a new generation of astronomers to work at the University Observatory. The death of Frank Bellamy freed up funding for a post, and in 1937 the American astrophysicist Thornton Page became the new First Assistant. That same year, Madge Adam, Plaskett's first research student at Oxford, became a research assistant at the observatory, initially studying plates of solar spectra originally obtained by Plaskett while he was still in Canada. She also became an assistant tutor at St Hugh's College, then a college for women students.

As part of its settlement with the Radcliffe trustees, the University gained funding for a 'Radcliffe Travelling Fellowship' in astronomy, which Plaskett used to attract talented young astronomers from abroad. The first of these was a Dutch astrophysicist, Herman Zanstra, and another postdoctoral researcher in the late 1930s was Reginald Victor Jones, who studied the infra-red spectrum of the Sun; he would soon leave astronomy to commence his famous work as a scientific intelligence officer in World War II.

Shortly after arriving at Oxford, Plaskett introduced a weekly colloquium at the Observatory, at which a research student or another junior researcher would talk through a published paper, and those present at the colloquium would then discuss it. Sometimes, new results were presented. True to Plaskett's ideal of observers and theorists working together, these colloquia were also attended by Milne and his research students.[41]

Plaskett also reformed the undergraduate teaching syllabus, which was still dominated by the old positional astronomy from the end of Turner's time, and he quickly introduced astrophysics to the undergraduate astronomy course. According to Madge Adam, these courses were very popular with undergraduates, who 'discovered the existence of the Observatory and kept coming!'.[42] But in 1946 this course was abolished, because Plaskett believed that those aspiring to become astronomers should read mathematics, physics, or both for their first degree. Astronomy instead became an optional subject in the Science Honour moderations.

Madge Adam (1912–2001) was the first research student in solar physics at the University Observatory. The first woman to achieve an Oxford First Class degree in physics, for many years she taught physics and astronomy at St Hugh's College.

In New College Plaskett was less involved in college business than Turner had been, although he served as sub-warden of the College from 1948 to 1949. However, he engaged more with University management than Turner, serving on Hebdomadal Council from 1946 to 1951 and becoming a Curator of the University Chest (its finance committee) from 1952 to 1956.

Plaskett was brought up by a generation of Canadians who still looked up to Britain as the 'home' country, which they felt bound to serve when it was in peril. In 1916 he had enthusiastically joined the Canadian army, and when war broke out again in 1939, he was as patriotic as any British citizen of his time, and joined the Territorial Army even before the war had started. But eventually, like many scientists of his time, he came to use his scientific skills in support of the war effort, spending 1940 to 1944 working for the Ministry of Aircraft Production which was based at the Aircraft and Armament Experimental Establishment at RAF Boscombe Down in Wiltshire. Projects that he worked on included minimizing the errors in sextants used for astronomical navigation aboard aircraft and investigating the deviations in the magnetic compasses on these aircraft. Madge Adam served as acting director of the University Observatory throughout Plaskett's wartime absence, although Plaskett managed to return to Oxford from Boscombe Down every weekend, and for meetings of the Observatory's Board of Visitors.

Like Turner before him, Plaskett held important positions in the Royal Astronomical Society, serving as one of its two secretaries from 1937 to 1940, and its president from 1945 to 1947. But perhaps his greatest service to the Society was in protecting it from enemy action during the war years. In the summer of 1939 Plaskett arranged for its most valuable archives and other possessions to be moved for safe keeping to the Oxford University Observatory, where

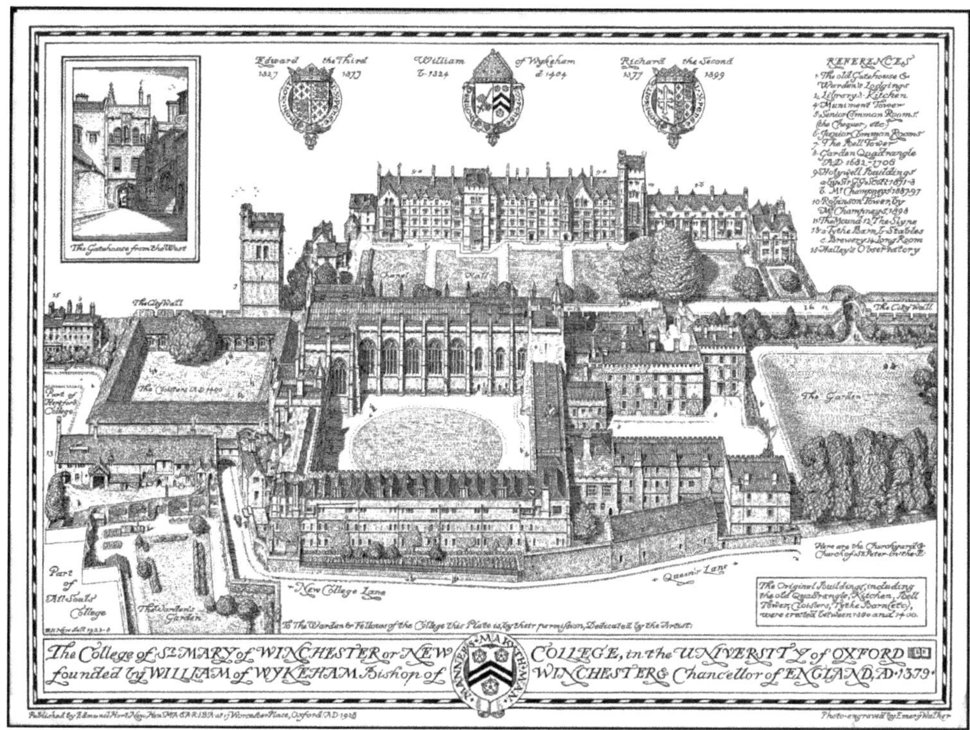

A 1920s engraving of New College, Oxford, by Edmond Hort New.

they remained throughout the war years. He also enabled its full-time assistant secretary, Kathleen Williams, to run the society from the relative safety of Oxford from the summer of 1939 until the spring of 1942; she worked in Plaskett's office while Plaskett was away on his wartime duties.[43]

Plaskett's long-held insistence that observers and theorists should work in close physical proximity led him to one serious mistake. As World War II drew to a close, he began to make the case for a large telescope to be built in Britain and made available for the common use of university astronomers from across the country. Airing these arguments publicly during his first presidential address to the Royal Astronomical Society on 8 February 1946, he swept away the idea that Britain's cloudy weather and unsteady atmosphere did not justify the use of large telescopes, using arguments such as:[44]

> The very fact that [William] Herschel, Lord Rosse and [Andrew] Common successively and successfully used apertures of 48, 72 and 60 inches indeed suggests that the seeing here must be at least comparable with that prevailing in other places where large instruments are used.

However, he failed to point out that Herschel, Rosse, and Common never used their giant telescopes on a regular basis, precisely because the observing conditions were seldom good enough for such large apertures to be effective.

To make efficient use of clear intervals in the British skies, Plaskett stated the case for a dual-purpose telescope of a design never attempted before. He proposed the construction of a 74-inch reflector, with a corrector plate enabling it to be used as a Schmidt camera for wide-field photography on moonless nights. The corrector plate would be removable, so that the telescope could be converted to a conventional Cassegrain reflector for the spectroscopy of individual celestial objects when the Moon was in the sky. Plaskett proposed applying to the government for a grant of £100,000 to build such a telescope. The astronomical community enthusiastically agreed to support the project, and in the summer of 1946 an application to the government was made, through the Royal Society.

The tercentenary of the birth of Isaac Newton, delayed for four years by the war, was then being publicly celebrated, and the government agreed to announce funding for the project in time for the celebration; the telescope thus came to be called the 'Isaac Newton Telescope'. By this time the aperture of the proposed telescope had increased to 100 inches, possibly to give it greater prestige, and astronomers decided that it should be built at Herstmonceux Castle, the post-war home of the Royal Greenwich Observatory.

But after this dramatic funding announcement, the Isaac Newton Telescope project became mired in delay after delay, partly due to financial difficulties in the post-war era. Moreover, the second-hand 98-inch primary mirror acquired for the telescope proved to be faulty, and grinding it to the correct curve took several years, while the technical problems with building an interchangeable telescope of this size proved to be insurmountable.

The instrument was eventually built as a fairly conventional 98-inch Cassegrain reflector and was inaugurated by Queen Elizabeth II in 1967. But despite one early success (the discovery of what is now known to be a black hole, Cygnus X-1), observing conditions at Herstmonceux proved to be hopelessly inadequate, and by the early 1970s astronomers were seeking to move the telescope to a better site abroad. It was dismantled in 1979 and re-erected with a new mirror on La Palma in the Canary Islands, since when it has gone from strength to strength.[45]

After World War II Plaskett returned to his solar research at Oxford. He supervised the construction of a larger solar telescope, with an aperture of 51 centimetres and a focal length of 35 metres, which gave a more generous image scale than the original, and fed a spectrograph mounted in a stable basement 2.4 metres below ground. The whole assembly was mounted in a separate tower situated to the north of the east dome that contained Plaskett's original solar telescope, and planning permission had to be obtained for the new building. The telescope was launched in June 1954. Working from Oxford, Plaskett helped to discover regular solar oscillations; these form the basis of 'helioseismology', now a recognized way of probing the interior physical properties of the Sun.

During the post-war years, Plaskett supervised a number of research students and postdoctoral researchers who went on to become distinguished astronomers: Radcliffe travelling fellow Walter Stibbs became professor of astronomy at St Andrews University, Victor Clube became a prominent figure in Oxford University's astrophysics department and a well-known writer on the potential consequences of an asteroid impact on Earth, and doctoral student Jack Meadows became a professor of astronomy at Leicester University and a noted historian of astronomy.[46]

During Plaskett's time in the Savilian chair, the Oxford University Observatory became a powerful astrophysical research centre with observers and theorists vigorously exchanging ideas; indeed, it might seem to resemble an astrophysics department of the early 21st century, but there were striking differences. In particular, Plaskett was the sole author of nearly all his scientific papers – a dramatic contrast from current astronomical papers which typically have many authors – with one obituary reporting Plaskett's dislike of group activities.[47]

He also reportedly did not enjoy lecturing, although his students and colleagues all noted that his lectures were meticulously prepared and presented. Like Turner, he sought to bring astronomy to a wider audience, but rather than lecturing to the Workers' Educational Association he instigated 'open nights' on Saturday evenings during Michaelmas and Hilary terms, when members of the University and their guests could observe celestial objects through the 12-inch refractor which still resided in the west dome at the Observatory. Perhaps surprisingly, given his transatlantic background, he travelled little after starting at Oxford, seldom venturing outside the United Kingdom.

Harry Plaskett seems to have been happiest when pursuing his solar research at the University Observatory – as he was able to do long after his retirement in September 1960. His successor, Donald Blackwell, let him continue to use the solar telescopes at Oxford, in addition to carrying on with his research from home where he maintained his output of published papers on solar spectroscopy and motions up to the late 1970s. On 26 January 1980 he died in Oxford, his adopted city, at the age of 86.[48]

Donald Blackwell

There was initially much more continuity between Plaskett and his successor than after Turner's death in 1930. Prior to Plaskett's retirement, the future of astronomy in Oxford was again reviewed by a committee, which recommended that astronomy should continue to be studied at Oxford but with no major investment in equipment.

By the late 1950s the University Observatory was under attack from two directions. Possibly because of the wider economic problems of post-war Britain, the University repeatedly cut the Observatory's annual budget, and in 1958 these cuts amounted to 20%. The Observatory was also meeting increasing competition for space from other departments wishing to erect buildings in the locality of South Parks Road. At a meeting on 8 February 1960, the heads of the science departments grumbled about 'the uneconomical use of the Observatory site and the land surrounding it'. Yet Plaskett stood firmly against either placing the telescopes on top of another building, or putting up large buildings nearby that might obscure their view of the Sun or otherwise interfere with solar observation. The possibility was raised of moving the Observatory to somewhere nearby, such as to Wytham to the west of Oxford or to Norham Gardens on the other side of the University Parks, but all agreed that the cost would be prohibitive unless the Observatory site were to become so valuable that its sale would pay for the cost of removal.

The question of the observatory site was still unresolved in October 1960 when Donald Blackwell was elected as the new Savilian professor of astronomy at the age of 39. Because of the financial constraints and the uncertainty over the site, the committee on the future of astronomy in Oxford recommended that Plaskett's successor should be either a solar observer (like Plaskett) or a theorist. The electors went for the first option.[49]

Donald Eustace Blackwell was born on 27 May 1921 in north London, the son of a civil servant, and was educated at the Merchant Taylors' School, a long-established private school in Hertfordshire. From there, he went to Sidney Sussex College, Cambridge, where he took a Double First in the Natural Sciences Tripos. After war service at Farnborough, he returned to Cambridge to begin a PhD degree. In 1947 he was awarded the Isaac Newton studentship for graduate students in astronomy, but wrote his thesis on infra-red spectroscopy at Cambridge's Institute of Colloid Science. By entering astronomy from a slightly different direction, Blackwell resembled his Cambridge near-contemporary David Dewhirst who began a doctorate in metallurgy before moving into observational astronomy, and the French

Donald Eustace Blackwell (1921–2010).

solar astronomer Bernard Lyot who studied engineering before finding employment in astronomy. In 1951 Blackwell married Nora Louise Carlton and would have two sons and two daughters.[50]

As at Oxford, Cambridge University in the mid-20th century had an active solar physics observatory, near to the University's old observatory in the grounds of what is now the Institute of Astronomy, and it was here in 1950 that Blackwell began his first professional appointment as its assistant director. Some of this observatory's most important work was carried out away from Cambridge, and in the era before space-based telescopes it was often necessary to go on expeditions to study aspects of the Sun, such as the corona, the granulation, and parts of the solar spectrum that were invisible at ground level. Options available to astronomers included travelling to observe a solar eclipse, or observing the uneclipsed Sun from aircraft, high-altitude balloons, or high mountain sites.

By 1950 Britain had a long tradition of sending official expeditions to observe solar eclipses. These were organized by the Joint Permanent Eclipse Committee of the Royal Society and the Royal Astronomical Society, which since 1894 had coordinated eclipse expeditions, applying for government funding through the Royal Society and seeing the scientific results from the expeditions through to publication.

While at Cambridge Donald Blackwell took part in expeditions to eclipses – in 1952 when he made infra-red measurements of the corona during an eclipse in Khartoum, in 1954 when he carried out photometry of the outer corona through the open door of a high-flying aircraft, and in 1955 when he attempted similar observations during an eclipse over the Pacific Ocean but was foiled by problems with the flying boat on which he was travelling. In 1958 he led an expedition to a site more than 5000 metres high on Mount Chacaltaya in the Bolivian Andes, to study the zodiacal light, a faint glow in the sky along the ecliptic (the Sun's apparent path in the sky) caused by dust orbiting the Sun and recognized by then to be a component of the corona. Also in the 1950s, along with David Dewhirst and the French astronomer Audouin Dollfus, he took photographs of the solar granulation from telescopes on high-altitude balloons, above the blurring effects of the Earth's lower atmosphere.[51]

When Blackwell took over as Savilian professor, the programme of work at the University Observatory was still heavily dominated by solar research, as it had been throughout Plaskett's tenure. Blackwell continued with his own solar work in his early years at Oxford, going on a second zodiacal light expedition to Bolivia in 1961 and observing another eclipse by air from northern Canada two years later. In 1965 he tried to observe an eclipse from Manuae Island, an atoll in the South Pacific, but in a classic instance of the weather's ability to ruin observations at ground level, a small cloud moved directly in front of the Sun during the eclipse.[52]

Astrophysics

By the early 1960s Donald Blackwell had realized that astronomy was changing fast, with the increasing use of large telescopes on remote mountain-tops, the rapid development of radio astronomy, and the possibilities opened by telescopes mounted on rockets and orbiting satellites. On 12 June 1962, while accepting that Oxford astronomy had to keep up with the times if it were to remain a credible University department and attract financial support, he asserted in his inaugural lecture as Savilian professor that modern astronomy was simply a branch of physics, with its use of the same techniques and apparatus, and should be approached as such. Desiring to do away with the University Observatory's Victorian image, he wished the Observatory to be renamed the 'Department of Astrophysics', and by 1964 this had happened when its annual reports to the Royal Astronomical Society began to be headed as such. In 1961 Blackwell negotiated a compromise concerning the Observatory's future, whereby it would stay on its existing site for fifteen years, with local building restrictions remaining in place to protect its view of the Sun. When this period expired in 1976, the Observatory was to move only when a better site became available.[53]

In the event, the issue of a new site in the Oxford area became irrelevant. By 1976 the Observatory was being used infrequently for direct observation, while the observing programme at Oxford had diversified beyond recognition, as Blackwell had anticipated in his inaugural lecture – indeed, he had expressed a desire for such change to happen. Believing that a university research department 'must lead and not be content to fill in the trivial gaps in our knowledge', Blackwell wanted to see departments like his own embrace new techniques, such as using Oxford equipment on large telescopes at better sites overseas, and observing from space with telescopes mounted on rockets and satellites. Space-based observations would allow the study of celestial objects at wavelengths blocked by the Earth's atmosphere, in addition to observations at optical wavelengths unaffected by atmospheric effects.[54]

Blackwell soon turned these ambitions into reality. By 1970, in addition to solar research, the department of astrophysics was carrying out work in stellar astrophysics (in which Geoffrey Smith, who also provided the backbone of the astronomy teaching courses, was a notable figure) and the properties of galaxies (under the direction of John Peach).[55] The department had also branched out into theoretical astrophysics and cosmology, especially after the arrival in October 1970 of Dennis Sciama, a Cambridge theoretical physicist who had done much to revive interest in cosmology among physicists. Sciama and his group worked in a large prefabricated wooden

A drawing of Donald Blackwell by the American artist Marc Winer.

hut in the Observatory grounds, which became known as the 'theory hut',[56] but sadly University politics and cuts to Science Research Council funding led to this group's break-up in the early 1980s.

Remote observing

By this time, as anticipated in Blackwell's inaugural lecture, the ever-broadening research programme was making increasing use of telescopes overseas, including the Anglo-Australian Telescope and the UK Schmidt Telescope in Australia. Blackwell's own research on the effective temperatures of stars was carried out using observations made at infrared and visible wavelengths in the Canary Islands. Oxford's astrophysicists were also using data from space-based observatories, such as Phil Charles's work in X-ray astronomy using the High Energy Astronomy Observatory 1 and Einstein X-ray satellites. Also important in the area of high-energy astrophysics was Janet Drew, whose appointment in 1985 was the result of a successful bid, led by Blackwell, to the British government's 'New blood lectureships' scheme, and who brought a mixture of theoretical and modelling insights into this field. Increasingly, researchers in Oxford were collaborating with workers in other universities, and by the time of Blackwell's retirement in September 1988 these trends had progressed further. Multi-institution collaborations had become the norm – for example, Oxford was one of six institutions that were analysing observations of the first naked-eye supernova in several hundred years, SN1987A in the Large Magellanic Cloud.[57]

Solar observations with Oxford's two solar telescopes continued throughout the 1970s and beyond, but Blackwell and some colleagues had also begun carrying out solar work abroad, from a tower in the Gornergrat hotel in the Swiss Alps, at an altitude of over 3100 metres.

The Warden and Fellows of New College, 1981; Donald Blackwell is at the right-hand end of the third row.

Assisted by David Petford, whom Blackwell had brought with him from Cambridge, he began to use photoelectric (rather than photographic) techniques to analyse the solar spectrum. Petford was famed in the department for his skills in engineering and computer programming, the latter becoming increasingly important in the 1970s.[58]

In the late 1960s Blackwell embarked on a major programme of laboratory studies to assist with the interpretation of solar and stellar spectra. He was interested in measuring the abundances of chemical elements in the spectra of the Sun and other stars, as shown by the strengths of the absorption lines in these spectra. But in order to understand how the strengths of absorption lines related to the chemical abundances, he needed a better understanding of the atomic transitions that caused these absorption lines.

To fulfil this need, he commissioned the building of a carbon-arc furnace that heated a half-metre-long tube filled with an inert gas to approximately 2000°C. This temperature was high enough to vaporize a sample of metal placed inside the tube, and for the metal to produce an observable absorption line when the tube was placed between a calibrated lamp and a spectrometer – in the same way that the solar photosphere absorbs light from brighter and denser layers below.

The furnace was installed in the basement of the tower that had formerly contained De la Rue's 13-inch reflector, by then known as the South Tower. It was powered by a very large transformer which was built so that in the event of a short circuit the core would jump downwards and not destroy its surroundings. A cooling tower had to be installed in the Observatory grounds to dissipate the energy. The resulting measurements led to new standard sets of transition characteristics for the iron group elements commonly found in the Sun and Sun-like stars, to a precision of 0.5%, an order of magnitude better than any measurements made

Built in the 1960s, the Denys Wilkinson building houses the astrophysics subdepartment of Oxford University's department of physics.

elsewhere. In view of these precision transition probability measurements, Blackwell's solar iron abundance determination from them was internationally adopted as 'the standard', until it became clear that a three-dimensional treatment of radiation transfer in a hydrodynamic model of the solar atmosphere was needed to bring all determinations of the solar iron abundance into agreement.

After his retirement in 1988 Blackwell continued to use the furnace, until the University's health and safety officers expressed concerns about someone in his late 60s using a 0.2 megawatt furnace on his own.[59]

Donald Blackwell served as president of the Royal Astronomical Society from 1973 to 1975, playing an active role at its meetings and often asking challenging questions of the speakers. However, some of his personality traits resembled those of Plaskett, in that he was quietly spoken and preferred to concentrate on his science.[60] Tony Lynas-Gray, whom Blackwell appointed to the department during his last year as Savilian professor, recalls that no amount of work was too much trouble for Blackwell when it came to eliminating sources of error from measurements – indeed, he would often repeat his experiments many times to ensure the highest level of precision.

Jon Godwin, a doctoral student and then a researcher in the department of astrophysics from 1973 to 1988, recalls that in these years 'there was still a lot of respect for one's elders' and that Blackwell 'was like the Headmaster, with whom one mainly associated on serious business, although he was a perfect gentleman and bore a kind heart'. He ran the department's business 'punctiliously, polishing the reports and balancing the books to perfection'.[61] Phil Charles, a faculty member at the time, similarly remembers Blackwell as being like a 'Victorian gentleman', a polite and kind person, who might have had difficulty surviving in the commercialized and managerial 21st-century university.[62]

Nevertheless, from his inaugural lecture in 1962 until the end of his time in the Savilian chair, Blackwell was committed to modernizing astronomy at Oxford. For example, Phil Charles recounts how Blackwell gave him the strongest support in developing the ISIS (Intermediate-dispersion Spectroscopic and Imaging System) spectrograph for the newly built

William Herschel Telescope on La Palma; this spectrograph had a lifetime of 35 years. Its successor, known as WEAVE (William Herschel telescope Enhanced Area Velocity Explorer), was also developed at Oxford, continuing Blackwell's tradition of building instrumentation for use on telescopes at premium sites overseas.[63]

Blackwell's tenure saw the end of the old University Observatory. Prior to his retirement, yet another committee on the future of astronomy in Oxford agreed that the astrophysics staff should move to more modern facilities in the nearby nuclear physics building on Keble Road (now the nuclear and astrophysics laboratory in the Denys Wilkinson building), which dated from the 1960s. The astrophysicists vacated the University Observatory in 1988.[64] In 1995 the benefactor Philip Wetton donated a 16-inch telescope that bears his name and is used for teaching and public outreach. Initially installed in the now-vacated observatory, it too was moved to a new dome on the Denys Wilkinson building in 2005.[65] The old observatory building is now used by the University's security services.

Donald Blackwell died on 3 December 2010.

Conclusion

The ninety-four years between Herbert Turner's election and Donald Blackwell's retirement as Savilian professor of astronomy should not be seen as a straightforward transition from positional astronomy to modern astrophysics.

Although Turner never contributed greatly to astrophysics, this was not due to any lack of interest in the newer branches of astronomy, but funding constraints and his commitments to the *Carte du Ciel* project worked against him. Turner was actually very forward-looking, and even something of a visionary, as illustrated by his suggestion in 1907 to move the Radcliffe Observatory to the Transvaal and his later support for Harold Knox-Shaw's scheme for moving it to South Africa.

At a first glance, Harry Plaskett's impact on Oxford astronomy was also forward-looking, in that he directed the Observatory towards cutting-edge solar physics and established a research culture in which observers and theorists worked together. But while these changes were important, Plaskett's regime was more conservative than it might appear. During his tenure the University Observatory mainly concentrated on solar physics, most papers were single-authored, and after 1932, with the exception of his wartime service, Plaskett worked almost entirely from Oxford.

Although Donald Blackwell's period of office might initially have seemed a continuation of the Plaskett era, in the sense that he was a solar physicist, he was arguably more revolutionary than Plaskett. Early in his tenure of the Savilian chair he expressed a wish for Oxford to branch out into different research areas, using remote and space-based observatories and observing in different wavelengths, and he then implemented this within ten years. By 1988, with such a wide-ranging programme now frequently carried out through collaborations with multiple institutions and resulting in multi-authored papers, the old University Observatory had been transformed into a department that increasingly resembled the 21st-century department of astrophysics.

Joe Silk on the cover of *The Scout* magazine in 1958.

CHAPTER 8

The golden age of cosmology

PEDRO G. FERREIRA

The late 20th and early 21st century became an extraordinary era for the Savilian chair and the astrophysics subdepartment at Oxford. It coincided with the golden age of physical cosmology, in which the study of the origin and evolution of the Universe was transformed from a speculative and data-starved backwater, looked upon with some suspicion by astronomers and physicists alike, to a precise, data-driven, and well-resourced field, a jewel in the crown of modern physics. By chance, the Savilian professors in post during that period, George Efstathiou (from 1988 to 1998) and Joseph ('Joe') Silk (from 1999 to 2010), were two of the leading figures in the genesis of the modern age of cosmology, instrumental in laying down the groundwork for the revolution that was to come.

Introduction

The period that spans the tenure of the Savilian chair by George Efstathiou and Joe Silk also reflects and embodies a social and economic transformation in the UK. Both of these individuals came from non-academic, immigrant, and working-class backgrounds from London. Efstathiou, the son of Greek–Cypriot immigrants, and Silk, the son of Polish–Jewish immigrants, both went through Oxbridge and taught on the faculty of leading universities before arriving in Oxford.

Efstathiou's tenure at Oxford occurred during the tail end of the governments of Margaret Thatcher and John Major, during which English academia underwent a period of cutbacks and hardship, while Silk joined Oxford during the upsurge of the New Labour government at a time when academia was being rejuvenated and reinvigorated. Both were transformative figures for the astrophysics sub-department, overseeing expansion and the hiring of new faculty that would steer it firmly towards its current position as a leading centre for cosmology and astrophysics for the 21st century.

Pedro G. Ferreira, *The golden age of cosmology*. In: *Oxford's Savilian Professors of Astronomy*. Edited by: Robin Wilson and Steven Balbus, Oxford University Press. © Oxford University Press (2025). DOI: 10.1093/oso/9780198894292.003.0008

Cosmology and Oxford

Cosmology as a physical science came into being in the early 20th century. The ability to build bigger and better telescopes led astronomers to detect and describe ever fainter objects that were far more distant than had previously been conceived. The idea that there was more to the Universe than just our collection of stars, the Milky Way Galaxy, became firmly established only when the American astronomer Edwin Hubble showed that objects then vaguely classified as 'nebulae' were actually independent galaxies in their own right and located at distances much greater than the outer reaches of our own Milky Way Galaxy. In parallel (and crucially), Albert Einstein's proposal of the general theory of relativity, a universal theory of the gravitational force, appeared and could be used to construct a coherent and consistent mathematical model of the Universe. Einstein's theory implied that the Universe was evolving and, by the late 1920s, Edwin Hubble and his collaborators were able to show convincingly that it is expanding, with galaxies receding from one another at rates proportional to their separations.

While the expanding Universe, later dubbed the 'big bang theory', became well established and generally accepted, little progress was made into fleshing out the details of a fully working and physical model of the Universe from its inception. It was not until the 1960s that the impetus really took off to build a model that could incorporate the expanding Universe, while also addressing the creation and evolution of galaxies. The 1965 discovery of the relic microwave radiation left over from the big bang itself marked the beginning of an important new field of observational cosmology. On the mathematical side, the work of theorists started to combine many different strands of theoretical physics to make accurate predictions of the myriad of physical processes in the Universe. On the computational front, emerging digital technology allowed for the development of new algorithms to simulate different model universes directly. Last, but not least, was a systematic observational study of galaxies, their different types, how they evolved, and how they coalesced on the largest scales.[1]

Throughout the 20th century, scientists at Oxford had made contributions to foundational issues in general relativity with some bearing on cosmology. In the 1930s Edward Arthur Milne, the Rouse Ball professor of mathematics, devoted much of his work to understanding the mathematical structure of cosmology, while from the 1970s onwards Roger Penrose, a later Rouse Ball professor, contributed various insights to the origin and evolution of the Universe, with a particular interest in the initial conditions and the arrow of time. Throughout the 1970s Dennis Sciama of All Souls College led a large and influential group working on many aspects of cosmology, from the structure of galaxies and the evolution of the early Universe to the quantum properties of black holes.[2] And from the 1980s onwards James Binney, in the Theoretical Physics sub-department, developed a formidable reputation as an expert in galactic dynamics. But in general, Oxford did not play a major role in the development of physical cosmology until the arrival of George Efstathiou and Joseph Silk.

Joseph Silk: from London to Berkeley

Joe Silk was born in 1942 and, for the first few years of his life, grew up with his family in Paddington in London. At the age of 5 he moved to Tottenham, where he spent the rest of his childhood and youth. His father, a tailor, was born in London, the son of Jewish refugees with the name Sulkovich, fleeing from the pogroms on the Russian–Polish border. Family legend

Arthur Milne, Roger Penrose, and Dennis Sciama.

had it that his grandfather, a conscript in the Russian army, deserted and managed to board a boat for the US, but became so seasick that he disembarked in Southampton and finally settled in London. Silk's mother was 12 when she fled with her family from Poland to settle in London. His parents had met and married in London and, although of Jewish heritage, were only weakly observant of religious traditions, and were firmly assimilated into the English working class.

Naturally gifted in mathematics, Joe Silk passed the 11-plus examination and entered a local grammar school, Tottenham County School, where he flourished. Encouraged and inspired by his mathematics teacher, a Mr Philips, he sat the entrance exam to Cambridge a year early, but failed to get a scholarship. Planning to stay on at school for another year, he was surprised to be contacted by the Master of Clare College, directly offering him a place. Cambridge was slowly changing from an elitist enclave of privately educated men to an institution that took on students from a more diverse background. Clare College was at the forefront of the drive to accept students from less privileged backgrounds, and Silk was one of the first grammar school students to be accepted.

His move to Cambridge was the first occasion that he had lived away from home, and he used it to explore an England with which he was largely unfamiliar. On the weekends he would regularly go to London with his friends to explore the nightlife, and he often accompanied one of his best friends to the Lake District during the vacations. All of these diversions, as he later admitted, contributed to his performing poorly in his examinations, ultimately graduating with a second-class degree in mathematics.

Despite such distractions he was drawn into the orbit of such theoretical physicists as the fluid dynamicist George Batchelor, the field theorist John Polkinghorne, and the astrophysicists Donald Lynden-Bell, Leon Mestel, and Dennis Sciama. One day he stumbled into an advanced lecture by Sciama on Mach's principle, on how the distant stars might have an effect on the 'here and now', and how Einstein had tried to incorporate this into cosmology. That lecture, and his exposure to Sciama, would have a 'huge influence' (as he later put it) on the direction of his research

Unable to continue at Cambridge, Joe Silk briefly considered pursuing a career outside academia, and undertook some unsuccessful professional internships, one in an actuarial firm in London and another in a management course organized by Unilever. Deciding to continue

Joseph Silk (b. 1942).

his studies, he moved to Manchester where he enrolled in a diploma course, studying physics and astronomy. Whilst there he worked with the radio astronomer Roger Jennison, one of the founders of modern radio interferometry who, at the tail end of his career, was trying to devise new experiments to test the special theory of relativity. Jennison tasked Silk with developing the theory for an experiment he was working on, and was sufficiently impressed with his grasp of physics that he wrote a strong reference letter. As a result, Silk was awarded a scholarship from the precursor of the European Space Agency, then known as the European Space Research Organisation, to study for a PhD degree at Harvard University in the USA.

Silk arrived at Harvard in 1964 and chose to work with David Layzer, an eminent and idiosyncratic cosmologist who is mostly remembered for his pioneering papers on the cosmic energy budget and the role that gravity plays in it.[3] But when Silk got there, Layzer was championing his own theory that the Universe began in a cold dense state, as a highly structured crystal that had evolved into the cosmos we see today. One of his predictions was that, as the Universe expanded, the cracks in the crystalline structure would seed the formation of the galaxies. However, when Silk arrived Arno Penzias and Robert Wilson from Bell Laboratories had recently discovered the cosmic microwave background, the relic radiation with a current ambient temperature of a few degrees Kelvin, left over from the very early Universe.[4] This radiation was direct evidence that the early Universe had been very hot, completely counter to what Layzer was proposing. Despite Layzer's misgivings, he allowed Silk to study how galaxies might have emerged from such a hot beginning.

In his seminal paper, 'Cosmic black-body radiation and galaxy formation',[5] which would be the main output from his PhD studies at Harvard, Silk stated that 'the most outstanding

COSMIC BLACK-BODY RADIATION AND GALAXY FORMATION

JOSEPH SILK
Harvard College Observatory, Cambridge, Massachusetts
Received May 31, 1967; revised August 21, 1967

ABSTRACT

The possibility is examined that galaxies may have formed from an initial spectrum of primordial fluctuations of small amplitude (although not necessarily of small scale) With a cosmological interpretation of the 3° K microwave background radiation, it is found that fluctuations containing more than 10^{-4} $M_\odot$, but less than a certain critical size, would be optically thick at an epoch subsequent to the primordial fireball (i.e , $T < 10^{10}$ ° K), and would be damped out by radiative diffusion on a time scale short compared to the expansion time This critical size corresponds to the mass of a typical galaxy, and is of the order of 10^{11} $M_\odot$. The survival of primordial fluctuations to an epoch when galaxy formation may occur is found to imply an angular anisotropy in the 3° K radiation of between approximately 10″ and 30″, depending on the mean value taken for the present density of matter in the Universe.

An upper limit is also derived on the mass of a gravitationally bound condensation by considering the destabilizing tendency of the radiation A consequence of this result is that any large-scale (> 100 Mpc) anisotropy in the Universe would be consistent with a value for the mean density of matter at the present epoch of only about 10^{-29} g/cm^3 or less.

Joseph Silk's paper on cosmic black-body radiation, written while he was at Harvard University.

unsolved problem in any cosmological theory is the origin of galaxies'. He set out to propose, that under gravitational collapse, very small fluctuations in the primeval plasma in the early Universe would evolve and grow to form the embryos of galaxies. Working through the equations that describe the evolution of an interacting plasma of electrons, protons, and photons, he found that these initial fluctuations would indeed evolve, but that, when the electron and protons combine to form hydrogen, they would be damped due to diffusion of the photons through the partially recombined plasma. Remarkably he discovered that, at small scales, these fluctuations would be completely wiped out, leaving a typical scale which corresponded to, but was slightly larger than, the size of the galaxies that we observe today – in other words, the physics of the primordial plasma would conspire to generate the precursors of galaxies of just the right size. The damping mechanism that came into play here is now known as 'Silk damping' and has become one of the key ingredients of physical cosmology. It cemented Silk's reputation as an astrophysicist of note.

In 1968 Joe Silk married Wendy Kuhn and left Harvard for Cambridge University, for a postdoctoral research appointment to work with the English astronomer Fred Hoyle. Hoyle had been one of the pioneers of stellar nucleosynthesis, and was also one of the creators of the steady state theory of the Universe, a rival to the more established big bang theory. By 1968 observational evidence strongly disfavoured Hoyle's theory, but that did not deter him from working on it and promoting it. Silk was the 'big bang person' in Hoyle's group, and was allowed to do whatever he wished. He found the group atmosphere inspiring and, although he never collaborated directly with Hoyle, he did work with other postdoctoral fellows and visitors at Cambridge. As a result, his interests began to diversify and he started working on the interstellar medium, the X-ray background radiation, and nucleosynthesis.

Silk continued to pursue his new interests when, a year later, he moved to Princeton University to work with Lyman Spitzer in the astrophysics department. There, as with Hoyle in Cambridge, he was greatly influenced by Spitzer's style and interests, but did not collaborate directly with him, preferring to interact directly with his peers. He also had to confront the odd culture that then prevailed at Princeton, in which the departments of astrophysics and

Joe Silk (bottom row, second from left) attending the Woods Hole Oceanographic Institute Summer School of Astrophysical Hydrodynamics in 1966.

physics had relatively little contact. Consequently, he had almost no interactions with the gravity group led by Robert Dicke in physics, which had pioneered the technology to search for the cosmic microwave background, and where one of its members, Jim Peebles, was working on a detailed calculation of the evolution of fluctuations in the big bang theory that both rivalled and complemented Silk's.[6]

In 1970 Joe Silk took up a faculty position in astronomy at the University of California at Berkeley, won over by the winning pitch of the head of department, George Field. Berkeley was a perfect fit for Silk's interests and aspirations, as Field had similar interests, but there was also a keen interest in the early Universe and in the formation of galaxies. One of Silk's colleagues, Hyron Spinrad, was pushing the boundaries of how very distant galaxies could be discovered and studied, while the group led by Stuart Boyer was uncovering new vistas on the Universe by observing the cosmos with X-rays and (in subsequent years) deploying experiments to measure the cosmic microwave background in great detail.

By the time that he arrived at Berkeley, Silk had already developed two key aspects of his scientific persona: a large and diverse range of scientific interests, and an ability to collaborate widely with peers and young researchers alike. To begin with, he carried on the work he had begun as a doctoral student, studying how cosmic structures would evolve and how they would interact with cosmic radiation. Most notable was his work with his student Michael Wilson in the early 1980s, developing the mathematical techniques to predict the distribution of galaxies and the amplitude of the fluctuations in the cosmic microwave background radiation over a range of scales.

His short stints as a postdoctoral researcher in Cambridge and Princeton had led him into other fields, further developing his interests in more astrophysical processes. A particular strand of research that developed in full flow at Berkeley was finding how galaxies formed and evolved. This difficult problem, which involves many different aspects of physics from gravity to gas dynamics and from stars to black holes, regularly captured Silk's attention and developed productively. In particular, during his stay at Berkeley, he worked on the origin of dwarf galaxies, the fragmentation of gas clouds and star formation, the role of dissipation and feedback in galaxy formation, and the initial stellar mass function and the first stars, to name but a few. He also developed an interest in the growing variety of astronomical radiation backgrounds, going beyond the cosmic microwave background radiation to explore the X-ray and infra-red backgrounds, in order to try to understand whether they were cosmological in origin, or were more recent and perhaps local. He became, as one uncharitable referee called him, a member of the 'background brigade'.

A new interest, which came to dominate his later career, was in one of the great unanswered questions in cosmology: What is the 'dark matter' of which most of the Universe seems to be composed? By the mid-1970s, the idea began to take root that there were more to galaxies than atoms (or baryons, as their constituents are technically known). Radio and optical observations showed that the outskirts of galaxies were rotating far too quickly to be reined in by the gravitational force of the gas and stars that were actually observed. In parallel, new developments in theoretical particle physics, most notably the idea that all the forces in the Universe might be unified, led to a suite of proposals for possible dark matter candidates which might be experimentally observable. The challenge, then, was to devise ways for seeking the dark matter particles.

While at Berkeley, Silk was strongly supportive of attempts to build detectors, and more generally of experiments to search for the elusive dark matter particles. But one of his key contributions was to propose the *indirect* detection of dark matter. With Mark Srednicki, he realized that the most popular dark matter candidate was not actually completely dark, but generated high energy signals spanning cosmic rays and gamma rays; this created a whole new industry in the emerging field of particle astrophysics.[7] With Srednicki and Keith Olive, he explored the idea of looking at how dark matter particles might interact with the ordinary matter in stars, and the Sun in particular, generating high-energy fluxes of familiar 'standard model' particles, such as neutrinos. Furthermore, Silk and his co-workers found that dark matter particles would also cluster around black holes and at the centres of galaxies, leading possibly to very distinct signatures. The work of Silk and his collaborators has contributed to a huge interest in constructing neutrino, gamma ray, and cosmic ray observatories.

Most of the work that Silk undertook was collaborative – indeed, as significant as his scientific achievements were his ability and track record as a mentor. During his time there, Berkeley became a mecca for young astronomers and physicists who were interested in theoretical astrophysics and cosmology, and over the decades Silk guided and supported some of the most influential thinkers in the field. Among them was a young George Efstathiou.

George Efstathiou: the journey to Oxford

George Petros Efstathiou was born in North London in 1955, into a family of Greek–Cypriot emigrés. His parents had moved to the UK in the late 1940s as part of the mass migration brought to reconstruct the British economy. They were both from farming families in Famagusta who arrived in the UK to become members of a wide Cypriot community of family and

A 3-year-old George Efstathiou with his parents.

friends. Efstathiou's father worked first as a waiter in a restaurant, and later as an assembly line worker at Thorn–EMI on the production line for radios and televisions. Their social circle was sufficiently tightly knit that his first language remained Greek until he and his brother started to attend school.

The Efstathiou family moved to Wood Green, where George first attended an infants school before progressing to Tottenham Grammar School, near White Hart Lane. This was in 1966, during Harold Wilson's government, when there was a concerted effort by the Labour government to convert grammar and selective schools into comprehensive schools that would take children of all abilities. Within a year of his arriving at the school, the entrance requirements were abandoned, resulting in an exodus of some of the best teachers. He found it difficult to integrate into the new regime and, as with other academically able children, was frequently bullied and physically assaulted by other students. As he recalls: 'I had been stabbed three times by the time I reached 16'.

By the time that he passed his O-levels (examinations that all students were required to take at age 16), Efstathiou found the situation intolerable. One of his physics teachers, Dick Yarrow, an Oxford graduate, saw his potential and suggested that he drop out of school and study on his own for his A-level examinations, to be taken just before university entrance. In the meantime, he was hired to work for Yarrow as a school laboratory technician until he was ready to take the entrance examination for Oxford. He passed the exam and was offered a place at Keble College to read physics.

As it had been for centuries, Oxford University was then one of the bastions of learning in the UK and the wider world, and yet for the young George Efstathiou it was a disappointment. For a start, he felt completely detached from the social life and alienated from most of his colleagues. As he recalls: 'I didn't fit in. It was a different social class'. Furthermore,

George Efstathiou with his school mentor, Dick Yarrow.

he found the physics degree poorly managed with an antiquated syllabus. Some of his personal tutors were ill prepared, and he often witnessed cases of students helping out the tutors to solve problems that they had been assigned for homework. As he recalls: 'It was an awful course . . .'

It was during his third year at Oxford that things took a turn for the better. For a start, he decided to step back and resort to his experience as an autodidact that had served him so well in his final years at school. Finding the lectures on atomic physics and quantum theory poorly explained and lacking depth, he decided to start relearning the fundamentals of modern physics from the fabled *Course of Theoretical Physics*, by the Soviet physicists Lev Landau and Evgeny Lifshitz. These books covered a wide range of topics, from quantum physics to classical field theory (including electromagnetism and general relativity), and he found that they gave him the integrated and coherent picture that the lectures lacked.

During that year he was given the possibility of writing an essay about a research topic. He chose 'The problem of missing mass', as the dark matter problem was then known. By this time Dennis Sciama, who had made such a great impression on Joe Silk at Cambridge, had moved to Oxford and was leading a research group, housed in prefabricated buildings in the University Parks. Efstathiou would spend time in its library, going over articles and discussing them with James Binney (then one of Sciama's graduate students), who later become a pioneer of galactic dynamics and an Oxford don himself. From Binney and his readings, Efstathiou's interest in contemporary research grew and he became aware of the still-embryonic field of physical cosmology. While finishing his BA degree in 1976, he became convinced that he wanted to proceed to a doctoral degree.

By this time Efstathiou had married his first wife, Janet Smart, whom he had met during his studies. They moved to Durham University where he was to be officially supervised by Richard ('Dick') Fong, a theoretical particle physicist who was then moving into cosmology and extra-galactic astronomy. Encouraged by the head of department, Arnold Wolfendale, Fong was building a group to focus on the large-scale structure of the Universe. He recruited Richard Ellis (who later became a major figure in the field of galaxy evolution) as a research assistant and Efstathiou as his first graduate student. While Ellis would work with Fong to develop Durham's observational programme to measure the statistics of the distribution of galaxies, Efstathiou was given the job of 'theorist' to develop the mathematical model which

Two astronomers: George Efstathiou and Richard Ellis in Crete.

they could then compare with data. He decided to focus on writing a cosmological 'N-body code', which was designed to calculate the evolution of a large number of galaxies under their mutual gravitational interactions.

Although there had been some work on cosmological N-body simulation,[8] it had used direct particle-to-particle methods that were laborious and computationally intensive, and were thereby limited in terms of their ability to model systems with large numbers of particles. Efstathiou decided to concentrate on methods which could use an average (or 'mesh-based') approach for particles that were farther apart, while saving direct summation (as in the particle-to-particle methods) only for particles that were close by and therefore required more accuracy. This was very novel for its time.

To learn the technique Efstathiou spent three months in Manchester working with the physicist Richard James, and a further month in Reading working with Jim Eastwood. He became the first astrophysical researcher to write a cosmological code, using what are known as 'P^3M (particle-particle-particle-mesh) methods' and enabling him to perform N-body simulations with tens of thousands of particles, a step change when compared with what had been undertaken until then.[9]

Efstathiou's self-directed work at Durham, carried out without any on-site technical supervision, was impressive enough for him to be noticed by the wider community. On completing his PhD degree in 1979, he was offered postdoctoral positions at Berkeley and at Cambridge. He was able to postpone the Cambridge position so as to spend a year working in Joe Silk's group at Berkeley, where he was part of an impressive group of postdoctoral researchers; Richard ('Dick') Bond, Mitchell ('Mitch') Begelman, and Bernard Carr were all at Berkeley at this time, and all went on to become renowned specialists in their fields. Efstathiou took advantage of this environment and started to diversify, working with a junior faculty member (Marc Davis) on galaxy surveys, Joe Silk on neutrino cosmology,[10] and Dick Bond on the cosmic microwave background. It was a remarkably productive year that would set him up with new avenues of research for his move to Cambridge in 1981. Three years later he was awarded a lectureship there, under the 'new blood' scheme established by the government.

The 'Gang of Four' (from left: Marc Davis, George Efstathiou, Simon White, and Carlos Frenk), with Joe Silk standing in the doorway.

At Cambridge, George Efstathiou pushed forward with what would become another of his main research interests: galaxy surveys. Towards the end of his PhD degree at Durham, the Anglo–Australian telescope had come online and, with Richard Ellis, Efstathiou had been awarded time on it for observations. One important project to come from this was the Anglo–Australian redshift survey, a three-dimensional catalogue of approximately 400 galaxies which could be analysed with the new statistical methods being adopted in cosmology.

It was also while in Cambridge that a colleague, Simon White, and a student, Carlos Frenk, along with their Berkeley collaborator Marc Davis, had become interested in testing the newly emerging dark matter paradigm in which the current matter content of the Universe consisted of 95% cold dark matter. Efstathiou began collaborating with them, using the N-body code that he had developed as a PhD student at Durham, and the group, dubbed the 'Gang of Four', was able to produce the first detailed simulations of what became known as the 'standard cold dark matter model', and then compare it with observations.[11] Their outcomes were promising, as they could mimic the observed large-scale structure in the Universe, but they were also troubling as the group found differences on smaller scales which were difficult to account for with the model. Nevertheless, their work ended up being the first systematic study of a complete model of the Universe which could, with some accuracy, describe its large-scale structure; this was an important achievement.

While at Berkeley, Efstathiou had started to develop the mathematical machinery for predicting the fluctuations in the cosmic microwave background, using some of the techniques that Joe Silk and his student Michael Wilson had proposed.[12] While Wilson and Silk had focused on finding a prediction in a Universe with only baryons, Dick Bond and Efstathiou concentrated on the microwave background in a Universe that included dark matter. By 1987 they had developed a powerful numerical code, leading to the first detailed statistical prediction (the angular 'power spectrum') of the microwave radiation in a form that is currently used today.[13] Bond and Efstathiou's prediction set a target for experiments that attempted to measure anisotropies in the microwave background.

Efstathiou at Oxford: 'We were on top of everything'

In 1987 George Efstathiou became aware that the Savilian professorship at Oxford would soon become vacant. Donald Blackwell had been in post for 28 years, and during the 1980s there had been attempts to modernize the astrophysics subdepartment under the same new blood scheme that had allowed Efstathiou to be hired at Cambridge. Efstathiou recognized in these efforts an opportunity to build something new, and he put his name forward.

This was a bold move. He was just 32 years old and had only recently been appointed to a lectureship at Cambridge. Oxford was caught by surprise at such a young candidate, but although it was not customary to do so, the panel took the unusual decision to interview him for the statutory chair. Efstathiou did not hold back, declaring up front that he was interested in building up the department and modernizing it, and to do so he would need to hire at least two new lecturers. Oxford had yet to enter the modern mercenary world of high-powered recruits, where incoming chairs could demand new posts and negotiate major changes or salaries. To Efstathiou, the interview seemed to have been a failure, with a clear resistance to his views and demands, and so he was surprised when he heard, a few weeks later, that Oxford wished to offer him the post, to start in 1988.

To bring Oxford astrophysics into the modern era, even before he arrived, George Efstathiou began to put together a plan to make Oxford University a centre for astronomical observations. Throughout the 1970s and 1980s the UK had a very strong international presence in observational astrophysics, comparable only to the US, and was well advanced when compared to Europe. It supported several telescopes in Australia and the Canary Isles, which gave UK astronomers unparalleled access to observational facilities, specifically with telescopes of up to four metres in diameter. In that spirit, and on viewing the US competition, the UK science community was looking ahead to establishing a new set of yet more powerful telescopes with 8-metre mirrors, and a call was sent out for a new centre to house the UK office for large telescopes (later to become known as the 'Gemini telescopes'). Efstathiou saw this as a unique opportunity to build Oxford's astrophysics, and he put in a bid to the Science Research Council to house the office at Oxford. He was successful. To manage the operation, Roger Davies (then at Kitt Peak National Observatory in Arizona) was recruited to Oxford to lead the new facility.

With the observational programme at Oxford fully under way, Efstathiou set out to reinvigorate the astrophysics sub-department. Working with Phil Charles, who became its head in 1994, his strategy was to broaden the expertise across the entire field, and he hired Patrick Roche to develop an instrumentation laboratory in infra-red astronomy, Martin Ward to work on active galactic nuclei and black holes more generally, Lance Miller to develop quasar surveys, Steve Rawlings to work on radio cosmology, and Philipp Podsiadlowski to work on stellar astrophysics. These jobs were mostly advertised as an open competition in any area of astrophysics, and in so doing the sub-department was able to grow and modernize.

At the same time as George Efstathiou was building up Oxford astrophysics, he was pushing ahead with his own research interests, putting together a young team of doctoral students and postdoctoral fellows who could focus primarily on what could be learnt from galaxy surveys. As he commented about that time:

> I think that was my happiest time scientifically, because we were on top of everything ... We were churning out a lot of stuff on physical cosmology.

And so, while at Oxford, Efstathiou started to reap the rewards of the galaxy surveys that he had spent time in developing and, armed with his theoretical tools, he published a prescient paper in 1990 that would foreshadow the revolution that swept cosmology a decade later.[14] In a letter to *Nature* with Steve Maddox and Will Sutherland, he revisited the results from a 1984 paper written by the 'Gang of Four', now comparing the predictions of his N-body simulations with a detailed study of the statistical behaviour gleaned from a survey of approximately two million galaxies.[15] They saw that if the model for the clustering of galaxies that emerged from his simulations were to match the amount of clustering actually seen in the data, then the expansion rate of the Universe (as measured by the Hubble constant) would have to be much lower than what all previous observations had seemed to indicate.

Remarkably, Efstathiou and his team found that if they included an additional parameter in their modelling – a so-called 'cosmological constant', first studied by Einstein for very different reasons – with the property that it was responsible for up to 80% of the total energy density of the Universe, then a reasonable fit to the galaxy clustering behaviour emerged, together with an acceptable value for the observed Hubble constant. This was a startling result, and it would take until the turn of the century for the cosmological constant, now defined more generally as 'dark energy', to be adopted by cosmologists. In 1998, by observing the distances and recession velocities of distant supernovas, astronomers found direct evidence for the presence of a cosmological constant; this discovery was later recognized by the award of the 2011 Nobel Prize for Physics to Saul Perlmutter, Adam Riess, and Brian Schmidt.

In 1992 a group of researchers, led by George Smoot at Berkeley, announced that NASA's Cosmic Background Explorer (*COBE*) satellite had detected, for the first time, tiny fluctuations in the cosmic microwave background radiation over very large scales.[16] The discovery of these fluctuations launched a golden age of cosmology, because it gave a direct link between the early and late Universe and could be used to test the predictions for the growth of structure that Silk and Efstathiou had earlier proposed.

For Efstathiou this opened up a whole new window on the Universe, as he could take the earlier calculations he had produced with Dick Bond and rule out classes of models that were not in agreement with the *COBE* data, while also thinking about using future measurements of this relic radiation from the birth of the Universe. And so, in a collaboration with a team in Europe, he set out to propose, a new satellite mission that would be able to map out the fluctuations in the cosmic microwave background with yet greater precision. Efstathiou's role was to contribute to the scientific case for such a mission, taking into consideration all the possible complications that might arise from other 'contaminating' sources of radiation in the Universe. As a result, a European Space Agency mission, the Planck spacecraft, was launched in 2009.[17]

In the same vein, Efstathiou set to work on building the scientific case for the next generation of galaxy surveys. With the UK and Australian observational cosmologists, he proposed a new survey that would use the 2-degree field (2df) facility based at the Anglo–Australian

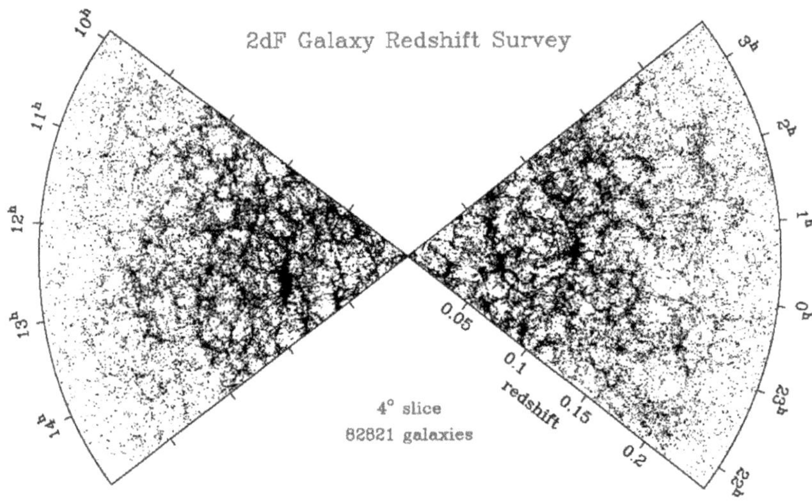

A diagram from the 2df survey.

observatory. This survey would consist of about a quarter of a million galaxies, selected from the Automated Plate Measurement galaxy catalogue on which he had worked. The observations for the 2df survey were taken between 1997 and 2002 and were finally made public in 2003, by which time Efstathiou had left Oxford.[18]

After a decade in Oxford, during which he had achieved what he had set out to do – to develop the subdepartment of astrophysics into a modern centre for observational astrophysics – Efstathiou felt that the winds were turning. The new head of department was less enthusiastic about growing the size of the sub-department than was Efstathiou, who felt that his ability to continue to develop it was being curtailed. Moreover, he was spending more time on managing the department than on his science:

> I found myself going through spreadsheets with pensions at 11:30 PM in my office in the department on a Saturday while I could hear people enjoying themselves in the streets of Oxford on a summer evening. There was something wrong.

Furthermore, his marriage had broken down. This was a further incentive to move elsewhere. The natural place for Efstathiou was to head back to Cambridge.

Joe Silk: leaving Berkeley for Oxford

By the late 1980s Joe Silk was, in many ways, at his scientific peak. Berkeley had been awarded a grant to house a new centre, the Center for Particle Astrophysics (CfPA), which could focus on topics that were dear to his heart: direct detection experiments for dark matter, automated supernova searches to do cosmology, and the development of techniques for directly observing the cosmic microwave background. Silk was given the brief to run the theoretical wing of the CfPA. He had developed a knack for hiring bright young cosmologists who would join an intellectually vibrant community working on any number of cutting-edge topics in cosmology.

George Efstathiou at his leaving party from the Savilian chair.

It was the work of his colleague and future Nobel Laureate George Smoot that, with the development of the *COBE* spacecraft, gave a dramatic new direction to his research in the cosmic microwave background, and set the scene for the developments of the 1990s. By then, Silk had accrued a large and formidable group of PhD students and postdoctoral fellows, covering all aspects of the cosmic microwave background, from technical data analysis to more speculative theoretical investigations. His gift for fostering a collaborative environment made Berkeley a major centre for research on the cosmic microwave background.

By the end of the 1990s, circumstances had changed for Silk at Berkeley. Following the COBE discovery and a period of intense collaborative creativity, he began to consider leaving the job that he had held for almost thirty years. His marriage had broken down and, during a sabbatical year in Paris, he met his future wife, Jaqueline Riffault, whom he would marry in 2001. He was unenthused by his teaching duties at Berkeley and, for the first time, felt unsupported within the department. So, when approached by the University of Oxford to be considered for the Savilian Chair, he saw an opportunity to start afresh and bring to Oxford his formidable experience in running a research group.

Whereas George Efstathiou had brought the astrophysics subdepartment up to speed, developing a strong instrumentation programme and a powerful array of observers, Silk felt that he could increase the focus on modern overlooked topics such as dark matter, dark energy, the early Universe, and numerical modelling of galaxy formations and evolution. These were topics that had been at the heart of his research for the previous three decades, and which he knew how to nurture. Efstathiou had been instrumental to the development of these topics in his research, but he had made no departmental hirings to sustain them on his departure. Silk saw this as an opportunity to make his own mark.

Given Joe Silk's reputation and seniority there was no need for an interview and, after a protracted negotiation about conditions, he arrived at Oxford in January 1999. One of the conditions that he placed was that he would be able to hire two new faculty members in areas of his choosing. The first of these would be in large-scale structure and the cosmic microwave background, with a focus on theory, and he appointed Pedro Ferreira to this role, starting in January 2000. The second post would be in numerical galaxy evolution and formation, and for this role he appointed Greg Bryan, who remained from 2001 to 2004, and then Adrianne Slyz along with Julien Devriendt, who started in the post from 2005. This combination was to lead Oxford over the next decade into becoming one of the key players in large-scale numerical simulations of galaxy formation.

Throughout his tenure at Oxford, Silk spent four days per week in the sub-department and three days in Paris, where he lived with his new wife Jaqueline. Associated with New College, he enjoyed collegiate life, regularly going in for dinners and becoming involved in other senior appointments in the college. In the sub-department, he set about creating the infrastructure for growing the areas that were of interest to him. He was able to secure stable funding for a new theoretical cosmology group in Oxford, and also led a successful bid for funding a European network of universities working on the cosmic microwave background (aptly named CMBNET), while simultaneously obtaining support from the Leverhulme foundation for supporting work on dark matter over a five-year period.

One of his major achievements, during the first few years, was the creation of a new institute for particle astrophysics and cosmology. In 2003 the philanthropist Adrian Beecroft offered to redevelop a wing of the Denys Wilkinson building, and to support fellowships in what would become known as the Beecroft Institute of Particle Astrophysics and Cosmology, more commonly known as the BIPAC; this institution, in its final incarnation, could accommodate up to twenty postdoctoral fellows and a number of visiting researchers.

During Silk's tenure at Oxford, the astrophysics subdepartment grew to become one of the largest astronomy departments in the UK. Research was strengthened with the new hirings, and through the efforts of Steve Rawlings, one of the faculty members appointed during George Efstathiou's tenure, Oxford became a major player in the design and construction of the Square Kilometre Array (SKA), the most ambitious radio telescope facility to date. Silk also recruited Chris Lintott as a postdoctoral fellow to the BIPAC, and helped to launch his subsequent career with another Leverhulme grant on citizen science. Lintott went on to create and lead Galaxy zoo and the Zooniverse, one of the most successful citizen science programmes to date.[19]

At Oxford, Joe Silk could concentrate on what had become his main interests towards the end of his time at Berkeley, and the quest for dark matter became one of the overriding themes of interest. While he did propose new and exotic models of dark matter, he also co-authored a major review article with two of his young BIPAC collaborators, Gianfranco Bertone and Dan Hooper, covering all possible particle candidates of dark matter.[20]

His main interest, however, was in dark matter detection. With his long-time collaborator Paolo Gondolo, he explored the possibility of detecting annihilation signals from dark matter at the galactic centre,[21] an idea that became central when a satellite (the Wilkinson Microwave Anisotropy Probe, or WMAP) detected a 'haze' in the Milky Way. He argued that dark matter could be clustered around black holes, leading to what have been dubbed 'dark matter' spikes, and with Max Bañados and Stephen West he proposed the possibility that particle collisions around spinning black holes would be greatly boosted, leading to the production of ultra-energetic particles.[22]

The interplay between high energy astrophysics and galaxy formation also took centre stage as Silk continued to pursue his interest in the roles that active galactic nuclei and black holes could play in the origin and evolution of galaxies and in the history of star formation and the interstellar medium. His ideas and work fuelled some of the research in the numerical galaxy formation group in the astrophysics sub-department, which tried to explore the ideas in more detailed and complex settings. His work on how these physical processes (and in particular the role of active galactic nuclei) influenced elliptical galaxies led to a surge of interest in how to incorporate these ideas into detailed models of galaxy formation.[23]

During his stay at Oxford, Silk added a new string to his bow. Throughout his career, he had had an interest in stepping back from the way that the field was evolving, in order to comment on more esoteric and metaphysical aspects of cosmology. Towards the end of his tenure in Oxford, he was awarded a large grant from the Templeton foundation to support meetings and workshops on 'Establishing the philosophy of cosmology', and subsequently, 'Fine-tuning in cosmology'. As a result, he was able to organize a series of meetings and debates with (and between) some of the most high-profile thinkers in cosmology and (more generally) in physics and astrophysics.[24]

In 2011 Joe Silk reached the retirement age at Oxford and decided to move full time to Paris, where he had been based at weekends throughout his tenure as a Savilian professor. He left a transformed astrophysics sub-department with an impressively broad range of expertise.

George Efstathiou: back to Cambridge

In 1998, as Joe Silk was being offered the Savilian chair at Oxford, George Efstathiou, recently married to Yvonne Nobis, took up the post of the 1909 Professorship of Physics at the Institute of Astronomy in Cambridge. He was to remain in that post for the next twenty-five years, until his retirement in 2023.

On moving to Cambridge at the age of 42, he decided against trying to repeat the feat that he accomplished at Oxford, where he had built a group of young people who could work on his projects and ideas. Instead, he returned to working in the way that he knew best and enjoyed most: on his own with his own ideas, and sporadically collaborating with others whenever this took his fancy: 'I became a postdoc again', he recalls. He was still involved in a series of observational projects, analysing data from a number of different astronomical surveys. But he also embarked on a period of smaller projects, focusing on galaxy formation and the intergalactic medium where he could explore what his friend and collaborator Dick Bond had dubbed 'gastrophysics'.

Cambridge would also provide another possibility for something he enjoyed: reshaping a department. In the first decade of the new century he was approached with the possibility of a donation from the Kavli foundation[25] to create a new institute which would bring together research in cosmology, which until then had been distributed between three Cambridge locations: the Institute of Astronomy, the Department of Applied Mathematics and Theoretical Physics (DAMTP), and the Cavendish Laboratory. Each of these had, at one time or another, hosted remarkable figures – Antony Hewish and Martin Ryle at the Cavendish, Fred Hoyle and

Martin Rees in the Institute, and Stephen Hawking at DAMTP – but surprisingly, and partly because of their geographic separation, they had often functioned as separate entities with little interaction between them.

The Kavli foundation considered a number of venues in the UK, among them Oxford with a bid led by Joe Silk, but Efstathiou's proposal won out. So from 2008 onwards, the infrastructure of cosmology research at Cambridge was gradually and irreversibly reshaped, with all the researchers moving into a new set of buildings attached to the Institute of Astronomy. The various Cambridge cosmologists would either move permanently or spend part of their time in the new Kavli Institute for Cosmology, Cambridge.

The endeavour that came to dominate Efstathiou's tenure at Cambridge was the Planck spacecraft mission, for which he had written the scientific case while still at Oxford. Initially, two competing missions were proposed to the European Space Agency, to map out the anisotropies of the cosmic microwave background radiation over the whole sky, down to arcminute resolution. Their proposers were asked to merge these into what ultimately became known as the Planck mission, named after one of the fathers of modern quantum theory, Max Planck, whose work was fundamental to the spectrum of the background radiation.

While Efstathiou was heavily involved in the design and preparation, he came into his own around 2009 when the Planck satellite was launched and data started coming in. He then followed the method of working that had served him well at school and university, during his PhD and postdoctoral periods, and then throughout his career: he developed (as far as possible, on his own) methods and techniques and undertook his own analysis of the data, often in parallel with other much larger teams within the collaboration. His efforts paid off, and the first data release from the Planck mission, with what were then the most precise constraints on the cosmological parameters, came from his personally managed pipeline.[26]

From the mid-2010s onwards, new wrinkles appeared in what seemed like a pristine and unfathomably successful cosmological model. While mild (and insignificant to many), two discrepancies began to emerge when data involving the early and late Universe were compared. Those from the earlier cosmic microwave background gave a marginally lower Hubble constant of about 67 kilometres per second per megaparsec, while those from more traditional methods using Cepheid stars pointed to a higher value of about 73 kilometres per second per megaparsec. In 2020 George Efstathiou took it upon himself to dig into the constraints coming from Cepheid stars and, in so doing, found some assumptions that, if loosened, would increase the uncertainty on the higher value and potentially lower the value of the Hubble constant.[27] Efstathiou's re-analysis opened up the possibility that, at least in some cases, the precision of the constraints was overestimated.

A milder puzzle that emerged related to the amount of clustering in the Universe. With the cosmic microwave background radiation, there seemed to be evidence for a higher amount of clustering than from surveys of the large-scale structure in one form or another. This discrepancy was most notable when one looked at the constraints that came from the weak gravitational lensing of distant galaxies. In the same way that he had re-analysed some of the Hubble constant data, Efstathiou set to work unpicking the analysis of the weak lensing data. In a couple of papers he showed what he thought to be troubling aspects in the Kilo–Degree Survey data (the most discrepant of the two data sets), but also how dramatically any conclusions

George Efstathiou (sixth from right) with the Planck Science team.

on clustering depended on the assumptions of how one modelled or calculated predictions in the standard model of cosmology.[28]

In 2023 George Efstathiou retired from his position at Cambridge, albeit remaining firmly embedded in the scientific and social life of the Kavli Institute and continuing to take an active role in UK science policy as a member of a number of influential committees.

Joseph Silk: the Paris years

Although Joe Silk had retired from Oxford in 2010, he maintained links with BIPAC and frequently visited there to run his programmes on the philosophy of cosmology and on fine-tuning in the physical universe. He was mostly based at the Institute of Astrophysics of Paris, where he had long maintained a close relationship. To these two affiliations he added Johns Hopkins University in Baltimore, where he was appointed a Homewood professor of physics and astronomy and where, on leaving Oxford, he would spend the fall semester every year. To top it off, he was awarded a prestigious advanced grant from the European Science Foundation, which allowed him to recruit a new cohort of postdoctoral fellows and graduate students, now based in Paris. And so started a new stage in his life.

Over the next decade Silk maintained his work on dark matter and galaxy formation, but two completely new interests emerged. The first of these arose from the discovery in 2015 of gravitational waves with the LIGO (Laser inferometer gravitational-wave observatory) detectors. The detection of such waves emitted by the merger of two black holes with a few dozen

Joseph Silk and his popular book *Back to the Moon*, published in 2022.

solar masses launched a new age of gravitational-wave astronomy and opened a new window into the Universe. He was naturally intrigued and enthused by the result and, with his doctoral students, set out to explore the possible consequences of the discovery and the potential of future gravitational wave experiments.

His second interest lay in the possibility of developing astronomical facilities on the far side of the Moon. Silk's original motivation was to produce a cheap and efficient method for deploying a very low frequency array of radio antennae that could map the 'dark ages' of the Universe, greatly increasing the depth with which it would be possible to map out the large-scale structure of the Universe. But as his interest in the Moon developed, he came to realize that there was a whole range of possible detectors that could be placed on the Moon, from gravitational-wave detectors to optical telescopes.[29] By the 2020s Silk had become an enthusiastic and articulate advocate for astronomy on the Moon, organizing workshops and publishing Op-Ed articles pushing for it.

Joe Silk was also involved with the popularization of his subject. As we saw in Chapter 2, Christopher Wren was appointed to the chair of astronomy at Gresham College in London, founded in 1597 for the purpose of giving free lectures to the general public. In 2015, Silk was appointed Gresham professor of astronomy, to be succeeded there by Katherine Blundell and Chris Lintott, both from Oxford's sub-department of astronomy.

Coda

At the time of writing, George Efstathiou and Joe Silk are still active in research and are sought-after speakers at conferences and workshops. They are both Fellows of the Royal Society and have been awarded prestigious prizes. Both have been awarded the Gold Medal of the Royal Astronomical Society and the Gruber Prize in cosmology. In addition, George Efstathiou has received the Hughes Prize of the Royal Society, the Maxwell Medal and Prize of the Institute of Physics, and the Dannie Heinemann Prize for astrophysics, while Joe Silk has been awarded the Bakerian Medal of the Royal Society, the Balzan Prize, and the Russell Lectureship of the American Astronomical Society.

Joseph Silk receiving the Gold Medal of the Royal Astronomical Society in 2008.

Both are recognized by their peers as exceptionally influential figures in the history of cosmology and in astrophysics more generally, and both are responsible for having transformed Oxford astrophysics into an internationally recognized centre of excellence.

Steven Balbus (b. 1953), on the occasion of his 2016 induction as a Fellow of the Royal Society.

CHAPTER 9

Interview with Steven Balbus

CHRIS LINTOTT

Steven Balbus is a leading theoretical astrophysicist, best known for the study of instabilities in rotating systems such as discs. He arrived as the Savilian professor in Oxford in 2012, following previous academic positions at the University of Virginia and at the École Normale Supérieure, Paris. In recent years, he has been awarded the Eddington Medal of the Royal Astronomical Society, and the Dirac Medal and Prize from the Institute of Physics, and in 2013 he shared the Shaw Prize in Astronomy with his collaborator John Hawley for determining how angular momentum transfer happens within accretion discs. He was the Savilian Professor of Astronomy from 2012 until October 2024, when he retired.

Early career

What led you to theoretical astrophysics? What was the state of play as you were becoming a PhD student?

In high school I had the math bug – my school let me work independently instead of going to classes – and so I went to MIT [the Massachusetts Institute of Technology], thinking that I would major in the subject. There, I found that what I was really interested in was problem solving, rather than abstract mathematics, and that led to my doing a double major in physics.

I'd always had a hobbyist's interest in astronomy, coming from my father who liked to observe the night sky. That gave me almost an emotional attachment to it; my father would point out Rigel and Sirius, Orion and Cassiopeia, and there wasn't a time in my life when I didn't know the major constellations. Every month he'd clip the sky at night column from the newspaper, complete with a star map.

I was also a reader of *Scientific American*, where Martin Gardner and Roger Penrose were talking about black holes, and Stephen Hawking in Cambridge was doing marvellous stuff showing that 'singularities', of which I had only a vague notion, appeared in various physical systems.

Chris Lintott, *Interview with Steven Balbus*. In: *Oxford's Savilian Professors of Astronomy*. Edited by: Robin Wilson and Steven Balbus, Oxford University Press. © Oxford University Press (2025). DOI: 10.1093/oso/9780198894292.003.0009

Some pictures from Steven Balbus's early years:
[*Above*] Steven as a toddler perusing a book with his physician father Theodore who was influential for Steven's childhood interest in science. Growing up in suburban Philadelphia, Steven was a competitive figure skater, pictured here with his skating partner Cynthia Ginkinger. [*Below*] Steven on the occasion of his 21st birthday, with father Theodore, mother Rita, and brother John. Steven, with full beard, in a formal portrait photograph as an undergraduate at MIT.

Putting this together, in the fall of 1974 it seemed as though a senior thesis in theoretical astrophysics would unify my interests, and it just so happened that the first binary pulsar (now called the 'Hulse–Taylor pulsar') had been discovered. There was great excitement, as this new object represented two compact objects orbiting very close to each other – the orbital period was just under eight hours. It therefore appeared that there was some possibility of seeing the effect of gravitational waves on the system, thanks to the precision of available pulsar timing.

It was like having nature's most accurate clock plonked down in a general relativity laboratory. So I set out to study this, working with my advisor Ken Brecher. He was something of an iconoclast, and thought the companion to the observed pulsar would turn out to be a white dwarf, because they were more common. In that case, tidal physics would be more important than gravitational radiation. So that was my first paper; a calculation done somewhat haphazardly, I would say now, but one which showed that tidal effects in this white dwarf–pulsar binary would indeed overwhelm the gravitational radiation energy losses.[1]

Happily, the companion was another neutron star, so we were completely wrong: the effects of gravitational radiation were in fact observed. But that was my entrée to astrophysics. I went to graduate school in Berkeley, and although I was still unsure about what subject to work in, I was staying with a X-ray astrophysicist named Roger Malina, until I found my own place. He invited me to the Astronomy Department's annual retreat at Yosemite National Park, which happened soon after I arrived. I got a lift with Suzanne McKee, who introduced me to her astrophysicist husband Chris McKee, with whom I got on well, and who became my PhD supervisor.

What was the style of PhD work at that time?

For a theorist at that time, it was a lonely struggle – you were often on your own, with general oversight from your supervisor. Chris would give me a problem, and I would meet with him weekly. Unlike a big observational project, there wasn't really a group; he would see people individually, but he wouldn't go through my work in detail.

Computers were big, bulky things. You didn't have one in your office, you had punch cards and stayed up late at night to get computer time. Looking back, it was a charmingly clunky way of trying to do problems, avoiding numerical approaches when you could.

The main problem I worked on for my PhD thesis was on the interstellar medium. People were trying to make sense of the new results from ultra-violet and X-ray astronomy; both were showing all kinds of unexpected radiation lines, both in emission and absorption, which had taken people by surprise. The then prevailing model of the interstellar medium, the Field–Goldsmith–Habing [FGH] model,[2] was mathematically very elegant, consisting of just two different phases – ionised hot gas at 10,000 K and cold clouds at 100 K. It was very simple and very elegant – but the interstellar medium turned out to be an awful mess: it didn't look anything like that, it was not elegant at all!

Chris and the well-known Princeton theorist Jerry Ostriker worked out a complicated but very influential three-phase model,[3] which included the energy input from supernovae – the first to incorporate what we would now call 'feedback'. It was clear that the interstellar medium was not a static place – not just a stable atmosphere for the galaxy – but was, in fact, dynamic. In between the hot and cold phases, you had clouds at 10,000 degrees, but it wasn't clear how you maintained this when it was embedded in a phase of gas at a million degrees. People thought that some of the lines that were being observed might come from evaporative interfaces between the phases, so I worked on that problem.

Part of the charm was that this was an unusual physics problem. Astrophysics is unusual because *where* things happen is important, and this particular problem turned out to be very complex as a result. It involves the physics of conduction and evaporation in what are very rarefied gases, but also plasma physics, and magnetic fields, with all of the associated potential instabilities that go with that. For me, it was a good introduction to theoretical work, because it impressed on me that research was more than just doing more complicated versions of the homework I'd been set as an undergraduate. You're doing stuff where you are completely on your own, where it isn't going to work out nicely, and where you need to learn how to cope with that.

In 2013 Steven Balbus and his co-worker John Hawley shared the prestigious Shaw Prize in Astronomy. This picture shows him receiving the award from Hong Kong's Chief Executive Leung Chun-ying.

What skills did you have to develop?

A good theorist needs to develop taste – the ability to decide which parts of a problem to care about. In astrophysics, you need to respect the received wisdom, but you cannot revere it. I've made progress on many problems because things which have been obvious to other people were not obvious to me. My motivation has always been to be obsessive about problems that I'm interested in, and to try to understand them deeply.

An example, the work that I'm best known for, which is a question about magnetic fields in rotating systems. Normally, if you try and stretch some gas, say, then any magnetic field will stretch with it, creating a tension which acts as a stabilizing force. But it wasn't obvious to me that things would be the same in a rotating system as in a static one, and I realized that almost no-one in astrophysics had bothered to attempt to do that problem quantitatively. And all of a sudden, it turned out that there's a new, completely unstable mode, now called the 'magneto-rotational instability'. It also turned out that others had, in fact, looked into different versions of this problem in other settings – Evgeny Velikhov and Subrahmanyan Chandrasekhar, for example, had studied the instability in the context of a magnetized cylindrical fluid flow in the laboratory.

Fortunately, I was working with John Hawley, one of the two or three numericists in the world who had a code that could study magnetized gases – a magnetohydrodynamic or 'MHD' code – back in the late 1980s, so we could make progress.[4] We were a good team. He had fantastic computational skills, and that's where he devoted his energies, while I could understand how to go from pencil-and-paper physics to something which could be simulated. At the time, people like him who could do technical work with computers didn't get the respect that they deserved, their work being very labour-intensive. That, happily, has now changed.

It's certainly true that computation has become more important.

It has, but pencil-and-paper calculations can still contribute. A recent example is in work led by my former student Andy Mummery, who has been interested in what happens in regions very close to a rotating black hole. We found a new, general mathematical form for how things are accreted in these very innermost regions, irrespective of the black hole spin – despite the fact that the Kerr metric which describes such a black hole has been known for more than sixty years.

There are always problems that people haven't yet looked at. I found that I liked to understand problems mathematically and that I'm not happy with half an explanation. At that time, there was just a limited number of people in astrophysics who brought powerful mathematical skills to the table, and so there were many problems that hadn't yet been worked on in depth.

You returned to MIT as a postdoctoral fellow?

Yes, I had a rather unrestricted position at MIT, with lots of freedom. With Scott Tremaine and Len Cowie present when I was there in the early 1980s, there was a lot of interest in the physics of discs. I worked with Len and a colleague who studied galactic structure, which is how my interest in that particular subject began, as I tried to understand how a disc responds to features such as spiral arms. While Frank Shu and C.-C. Lin had proposed that spiral arms were density waves, it became clear that this wasn't self-sustaining on its own. So how were long-lived spiral arms produced, what would be the response of the interstellar medium, and how would that affect star formation? These were all hot topics at the time.

I had colleagues at MIT, particularly Scott Tremaine with whom I spoke often, even though I published papers with Len Cowie. I had great respect for Scott's technical skills, his taste in problems, and his willingness to get his hands dirty. He told me that if you're doing a problem where you're having to make a series of approximations, you should think of a physical system – even if it's crazy – where these approximations are exact, and see whether the solution makes sense in that system.

Of course, as well as MIT there was the Harvard–Smithsonian Center for Astrophysics up the street, which all added up to a lot of talented people in the neighbourhood working on interesting problems. With several colleagues there, I became involved in working out what happens, from a dynamical point of view, if star formation that is triggered in a spiral arm itself triggers another round of star formation, followed then by supernovae and shock waves, as I tried to calculate how this sort of star formation would propagate.

It was a very happy time in my life. I felt greatly appreciated there, and this led to all sorts of new opportunities. One example which I'm very proud of, although it's not one of my best-known papers, followed on from my early interest in the problem of how clouds evaporate in the interstellar medium. It turns out that if you set up the problem in terms of a large ensemble of clouds in a hot medium, then what you're solving is the Laplace equation – the problem becomes mathematically identical to electrostatics. The temperature field is the analogue of the potential, acting as a sort of bath, and the expression for the individual mass loss rate is related by a constant factor to what would be the electrical capacitance. I felt like such a physicist, although not many people have cited that paper![5]

But in 1985 you moved to the University of Virginia as a faculty member?

Yes, that was the only assistant professorship that I was offered. At that time there were fewer job opportunities than there are now, and I was happy to go to Virginia. After the freedom of being a postdoc, it was also a bit of a shock, as I was pulled in many different directions, and I missed the very large number of different and stimulating people that I'd worked with at MIT and Princeton. But in talking with the excellent Virginia theorists Craig Sarazin and Roger Chevalier, I became interested in gas cooling flows found within clusters of galaxies, and in the problem of how to explain the resulting observed temperature profiles. I did some work in this field which got noticed, but the problems are still, to this day, not completely understood. But the best thing about Virginia was that John Hawley [mentioned above] turned up as an Assistant Professor a couple of years after I arrived. Hiring him positioned the department in just the right place for when computational astrophysics took off.

How did you and Hawley come by the problem you tackled together?

He had been at Caltech, working on discs. His simulations had been important in understanding the Papaloizou–Pringle instability,[6] which occurs in rotating systems like accretion discs. He showed me a paper where the authors studied a system that had a magnetic field threaded through the disc, which they had ignored as a leading approximation, because the field was weak. But I had already worked on magnetic fields, and I knew that it might be important to consider the field as present from the start, and that it could display all sorts of behaviour precisely because it *was* weak. In August 1990, I had been playing around with this problem.

It was a chaotic time personally, as the family had just moved house, but I eventually did the calculation and showed that such a disc would actually be very unstable in the presence of a weak magnetic field. John was on holiday at the time, but when he returned he asked me whether he had missed anything exciting. I told him both that we'd moved house and that I knew why accretion discs are unstable and probably turbulent. He raised an interested eyebrow, and I told him the answer was to put a magnetic field in – but that it wasn't cheating, because even a weak magnetic field would work. He went off to simulate it, and very quickly found the behaviour that the calculations had predicted.

I knew that we'd found something important, but I hadn't realized quite how important the simulations would turn out to be in convincing the theoretical community, who normally argue about everything, that the predicted behaviour was real. Simulations would also prove to be an

Steven and his wife, Caroline Terquem, relaxing during a break in a conference. Caroline, a fellow astrophysicist, is also a professor in the physics department at Oxford University.

important tool in understanding the phenomenon in real discs. Although there were detractors, within a year or so the community had quickly adopted our solutions.

What took you to Paris?

I had separated from my first wife in about 2000. My current wife Caroline Terquem, who is also a theoretical astrophysicist, was then working in Paris. Rather than lead a transatlantic existence, I applied for a job in astrophysical theory that had opened at the École Normale Supérieure in Paris, and I was offered the position. I was delighted to be there with Caroline and our baby daughter Elizabeth, but the one negative aspect was that it broke up my working relationship with John Hawley. We did continue to work together for a few years, until he became a Dean at the University of Virginia.

You broadened your scope of work again?

In Paris I became very interested in helioseismology, and in particular in how to explain the rotation pattern of the interior of the Sun. It's a complicated system, with turbulence due to convection as well as magnetic fields, but what we were trying to explain is actually very regular. From my work with discs, I had a sense of what to do with a rotating convective system, and

Steven, with astrophysicist colleagues Martin Rees (left, Baron Rees of Ludlow, former Astronomer Royal) and Andrew Fabian (right), attending a Royal Society conference on black holes in 2017.

was able to solve the fundamental physics equations analytically, in such a way that comparison with the data could be made. The results worked well.

What brought you to Oxford?

The person who suggested that I apply for the Savilian chair of astronomy was Alex Schekochihin, a plasma theorist at Merton College, who was keen to bring me to Oxford. I knew that I would have great students to work with; this was something that I had become used to at the ENS, and it was attractive to think about influencing the direction of such a great university. There was talk of additional junior positions in the department, although that proved to be more complicated than expected. I also knew that there was a very strong cosmology group that I could learn from, something that in retrospect I should have taken more advantage of!

I vividly remember the visit when I was shortlisted. I was interviewed in his lodgings by the then New College Warden, Curtis Price, and thought what an unbelievably beautiful place it was. I saw the children from New College School running through the porters' lodge gate, and thought of my son David who would be just starting school. It all seemed idyllic. We went to dinner, and I recognized that the real appeal of college life is that you're exposed on a regular basis to all kinds of fascinating and really bright people in different disciplines.

I should mention your interest in evolution.

I remember talking to colleagues back in Virginia about the problem of why the Sun and the Moon appear the same size in the sky. This apparent coincidence is related to the fact that the tidal forces which they induce are similar in strength, within a factor of 2 of each other. On the Earth there are two agents causing tides, with slightly differing frequencies, so they beat against each other, causing our spring tides and neap tides. These phenomena, which we take for granted, are actually a marvel of astrophysical fine tuning, which we might not expect on other planets.

The resulting variations in the tides mean that pools of water can be left far inland, and perhaps persist for months. If you're a fish, you had better learn navigational skills to flop around and perhaps return to the sea, or you're in trouble! The idea is that this unusual tidal variation might have been a key evolutionary driver in vertebrates coming out of the sea and onto the land;[7] this was a transition that seems to have happened only once in the Earth's history. The question I pursued was whether we could learn anything by returning to the reconstructions of continental positions at the time that this had happened and doing numerical simulations. It turns out that, although we don't know all the details, the tides would have been particularly large in regions where we find fossils of fish species transitioning to the land[8].

I mentioned this to the evolutionary biologist Richard Dawkins at lunch in college, and quickly found myself lecturing to the zoology department. Richard even describes the theory in the second edition of his book, *The Ancestor's Tale.*[9] It's a good example of the exchange of ideas that happens in college all the time.

Steven interviewing zoologist Richard Dawkins at the 2016 Starmus Festival, on the topic of life in the Universe.

Did you teach in college?

I did admissions! I lectured in the astronomy department too, of course, and met students that way – indeed, my latest research student was someone I sat next to at a New College physics dinner – and I also served on various committees.

There was also the case of a very gifted student who wasn't performing nearly as well as he should have been. There were issues with his family, as his father had asked to review his work before it was handed in. College told the father that he would be personally supervised by the Savilian Professor of Astronomy, and so for about a year I tutored this student – getting to see some physics problems I hadn't thought about for years – and it all worked out very well.

Within the department, I was the head of astrophysics for five years. This came more quickly than I had expected, and was a lot of work.

Finally, what advice would you have for your successors?

First, don't let the administration get you down; in Oxford, there's always a way around difficulties. Second, take advantage of college life, and get involved with the college in general. Finally, lecture when and where you can. Working with the students here is endlessly rewarding.

FURTHER READING, NOTES, AND REFERENCES

This section contains suggestions for further reading for each chapter, and detailed notes and references as indicated within the chapters.

Works cited in more than one chapter include:

John Fauvel, Raymond Flood, and Robin Wilson (eds), *Oxford Figures: Eight Centuries of the Mathematical Sciences*, Oxford University Press (2nd edn, 2013). [*Oxford Figures*]

William Poole and Christopher Skelton-Foord (eds), *Geometry and Astronomy in New College, Oxford: On the Quatercentenary of the Savilian Professorships 1619–2019*, New College Library & Archives (2019). [*New College*]

Oxford Dictionary of National Biography, Oxford University Press (2004 and later updates). [*ODNB*]

Companion works to this volume are:

Robin Wilson (ed.), *Oxford's Savilian Professors of Geometry: The First 400 Years*, Oxford University Press (2022). [*Savilian Geometry*]

Christopher D. Hollings and Mark McCartney (eds), *Oxford's Sedleian Professors of Natural Philosophy: The First 400 Years*, Oxford University Press (2023). [*Sedleian*]

Also useful are the various volumes of

The History of the University of Oxford, Clarendon Press, Oxford (1984–2000).

Useful writings on astronomical observatories include:

Roger Hutchins, *British University Observatories 1772–1939*, Ashgate (2008). [*Observatories*]

Jeffrey Burley and Kristina Plenderleith (eds), *A History of the Radcliffe Observatory Oxford: The Biography of a Building*, Green College, Oxford (2005), and especially the chapter by Roger Hutchins on 'Astronomical measurement at the Radcliffe Observatory 1773–1934' on pp. 63–101. [*Radcliffe*]

In general, for books published before 1900 only the places of publication are given, and for books published thereafter only the publishers are given.

CHAPTER 1

Further reading

General writings on the period covered in this chapter include the following:

John Fauvel, 'Eight centuries of mathematical traditions', *Oxford Figures*, pp. 3–34. [*Fauvel*]

Mordechai Feingold, *The Mathematician's Apprenticeship: Science, Universities and Society in England, 1560–1640*, Cambridge University Press (1984). [*Feingold*]

R. Goulding, *Defending Hypatia: Ramus, Savile, and the Renaissance Rediscovery of Mathematical History*, Archimedes, Springer (2010). [*Hypatia*]

William Poole, *Savilian Geometry*, Chapter 1. [*Poole*]

G. J. Toomer, *Eastern Wisedome and Learning: The Study of Arabic in Seventeenth-Century England*, Clarendon Press, Oxford (1996). [*Toomer*]

Notes and references

1. These statutes were similar to those that had been in place for over a century, while slightly increasing the amount of mathematics required of the student.
2. The main source here is *Feingold*, which sets out the personalities and connections in rich detail.
3. Savile's translation was made from the *editio princeps* of the *Almagest*, published in Basel in 1538, which includes these precise commentaries. His own annotated copy of this edition is in the Bodleian Library (Savile W.14), while his *Almagest* translation occupies MSS Savile 26–8.
4. Savile's heavily annotated copy of this 1533 edition has the Bodleian shelfmark Savile W.9. On his annotations and his early education in general, see Robert Goulding, 'Henry Savile reads his Euclid', *For the Sake of Learning: Essays in Honor of Anthony Grafton* (ed. A. Blair and A.-S. Goeing), Brill (2016), 780–97.
5. See Goulding (note 4), pp. 780–2.
6. G. H. Martin and J. R. L. Highfield, *A History of Merton College, Oxford*, Oxford University Press (1997), 184.
7. The first reference to Copernicus in Savile's lectures is in his 21st lecture, at MS Savile 29, fol. 90. In MS Savile 31, fol. 3, he compares the tables of mean solar motion of Ptolemy, Copernicus, and Erasmus Reinhold.
8. On the chronology of Savile's lectures, and their disruption by the plague, see *Hypatia*, pp. 187–9.
9. On Savile's tour, see *Feingold*, pp. 125–9; Robert B. Todd, 'Henry and Thomas Savile in Italy', *Bibliotheque d'humanisme et renaissance; travaux et documents* 58, no. 2 (1 January 1996), 439–44; Robert Goulding, 'Numbers and paths: Henry Savile's manuscript treatises on the Euclidean theory of proportion', *Reading Mathematics in Early Modern Europe* (ed. B. Wardhaugh, P. Beeley, and Y. Nasifoglu), Routledge (2020), 41–2.
10. See Robert Goulding, 'Henry Savile and the Tychonic world-system', *Journal of the Warburg and Courtauld Institutes* 58 (1995), 152–79.
11. See Goulding (note 9).

12. Jean-Louis Quantin, 'Du Chrysostome latin au Chrysostome grec: Une histoire europeenne (1588–1613)', *Chrysostomosbilder in 1600 Jahren: Facetten der Wirkungsgeschichte eines Kirchenvaters* (ed. M. Wallraff and R. Brändle), Arbeiten zur Kirchengeschichte, Walter de Gruyter (2008), 267–346.
13. Robert Goulding, 'Polemic in the margin: Henry Savile against Joseph Scaliger's quadrature of the circle', in *Scientia in margine: Études sur les marginalia dans les manuscrits scientifiques du Moyen Âge à la Renaissance*, Droz (2005), 241–59.
14. Petrus Ramus and Audomarus Talaeus, *Collectaneae praefationes, epistolae, orationes* (Marburg, 1599), 174–5.
15. Petrus Ramus, *Prooemium mathematicum* (Paris, 1567), 59; see also *Hypatia*, pp. 112–13.
16. R. Goulding, 'Testimonia Humanitatis: The early lectures of Henry Savile', *Sir Thomas Gresham and Gresham College: Studies in the Intellectual History of London in the Sixteenth and Seventeenth Centuries* (ed. F. Ames-Lewis), Ashgate (1999), 125–45.
17. See *Hypatia*, p. 182.
18. Strickland Gibson, *Statuta antiqua Universitatis Oxoniensis*, Clarendon Press, Oxford (1931), 535.
19. On the long practices of reading Euclid by the Savilian professors, see the chapter on Edward Barnard in Benjamin Wardhaugh, *Encounters with Euclid: How an Ancient Greek Geometry Text Shaped the World*, Princeton University Press (2021), 81–9.
20. See Tabitta van Nouhuys, *The Age of Two-Faced Janus: The Comets of 1577 and 1618 and the Decline of the Aristotelian World View in the Netherlands*, Brill's Studies in Intellectual History 89, Brill (1998).
21. MS Oxford, Corpus Christi Library 254, fols 88v–89.
22. See *Fauvel*, p. 20.
23. See Thomas Smith, *Vitae quorundam eruditissimorum et illustrium virorum*, London (1707), sig. Gg 1; *Toomer*, p. 72, states that Smith's biography of Bainbridge is 'slovenly'.
24. The comet had disappeared from sight by 16 December 1618, according to Bainbridge's account. An edition of the work appeared with the date 1618, presumably in the closing days of the year, and another appeared in 1619. Both came from the same press in London, but were sponsored by different printers.
25. The very first of these astronomers was Kepler, who observed one of the earlier comets in that same year; see Rienk Vermij and Paul Vieth, 'Confessionalization and comets. John Bainbridge on the comet of 1618', *Annals of Science* 79, no. 3 (July 2022), 279–91.
26. See Vermij and Vieth (note 25), p. 280.
27. For the history of this theory, see Peter Barker, 'The optical theory of comets from Apian to Kepler', *Physis* 30, no. 1 (1993), 1–26. Among those who published on the 1618 comet, Willebrord Snell, the pioneer in the theory of refraction, stands out for his *rejection* of this optical theory; see van Nouhuys (note 20), pp. 346–8.
28. See Vermij and Vieth (note 25), pp. 280–2.
29. See Vermij and Vieth (note 25), p. 289.
30. Cited in *Feingold*, p. 143.
31. See Robert B. Todd, 'Geminus and the ps.-Proclan *Sphaera*', *Catalogus translationum et commentariorum*, Vol. 8 (ed. V. Brown, J. Hankins, and R. A. Kaster), Catholic University of America Press (2003), 14, 19–23, and Todd (note 9), pp. 440–2.

32. Savile frequently referred to Geminus on geometry throughout the *Praelectiones*, always piling praise upon him, as *doctissimus* (p. 29), *elegantissimus* (p. 80 and several other places), and *acutissimus* (pp. 163, 216).
33. See Bernard R. Goldstein, 'The Arabic version of Ptolemy's *Planetary Hypotheses*', *Transactions of the American Philosophical Society* 54, no. 4 (1967), 3–55, and Elizabeth Hamm, 'Ptolemy's Planetary Theory: An English Translation of Book One, Part A of the Planetary Hypotheses with Introduction and Commentary', PhD thesis, University of Toronto, 2011, 5–7.
34. See Claudius Ptolemaeus, *Opera quae exstant omnia*, Vol. 2: *Opera astronomica minora* (ed. J. L. Heiberg), B. G. Teubner (1907); Hamm (note 33), p. 11; and N. M. Swerdlow, 'Ptolemy's scientific cosmology', *Theory, Evidence, Data: Themes from George E. Smith* (ed. M. Stan and C. Smeenk), Boston Studies in the Philosophy and History of Science 343, Springer (2023), 348. See also Anthony Grafton, 'Barrow as a scholar', *Before Newton: The Life and Times of Isaac Barrow* (ed. M. Feingold), Cambridge University Press (1990), 294, which notes that Otto Neugebauer recognized the value of Bainbridge's edition, even though some of his most brilliant emendations had been passed over by Heiberg.
35. See *Toomer*, pp. 74–5.
36. *Toomer*, p. 72.
37. *Toomer*, pp. 72–3.
38. He made a brilliant emendation of an ancient eclipse; see Grafton (note 34), pp. 294–5. Thomas Lydiat attended lectures of Bainbridge's on the calendar in 1625 and was upset by what he took to be criticism of his own work on the calendar, although he continued for years later to consult Bainbridge on issues of dating. See also *Feingold*, p. 149.
39. See *Feingold*, pp. 146–7.
40. On this episode, see Goulding (note 13).
41. On Scaliger and the calendar, see Anthony Grafton, *Joseph Scaliger: A Study in the History of Classical Scholarship*, Vol. 2: Historical Chronology, Oxford–Warburg Studies, Oxford University Press (1983).
42. See Grafton (note 34), p. 295.
43. 'Qui Scaligerum felicius correxit, Quam Scaliger Tempora'. See Thomas Smith, *Vitae quorundam eruditissimorum et illustrium virorum* (London, 1707), 13.
44. See *Toomer*, p. 75.
45. See *Poole*, pp. 16–17.
46. *Poole*, p. 19.
47. *Poole*, pp. 24–5.
48. See Francis Maddison's *ODNB* article on John Greaves.
49. Zur Shalev, 'The travel notebooks of John Greaves', *The Republic of Letters and the Levant* (2005), 80.
50. Zur Shalev, 'Measurer of all things: John Greaves (1602–1652), the great pyramid, and early modern metrology', *Journal of the History of Ideas* 63, no. 4 (2002), 560–1.
51. Shalev (note 49), p. 80.

CHAPTER 2

Further reading

General books on this period include

Nicholas Tyacke (ed.), *The History of the University of Oxford*, Vol. IV: *Seventeenth-Century Oxford*, Clarendon Press, Oxford (1997). [*Tyacke*]
Mordechai Feingold, *The Mathematicians' Apprenticeship: Science, Universities and Society in England, 1560–1640*, Cambridge University Press (1984). [*Feingold*]

For a description of the intellectual life of Wadham College when Ward and Wren were there, see

C. S. L. Davies, 'Wilkins at Wadham', in William Poole (ed.), *John Wilkins (1614–1672): New Essays*, Brill (2017), 37–65.
C. S. L. Davies, 'The youth and education of Christopher Wren', *English Historical Review* 123 (no. 501) (April 2008). [*Davies*]

Biographical material can be found in

Walter Pope, *The Life of Seth [Ward]: Lord Bishop of Salisbury* (ed. J. B. Bamborough), published for the Luttrell Society by Basil Blackwell (1961), originally published in 1697. [*Pope*]
John Aubrey, *Brief Lives: with An Apparatus for the Lives of our English Mathematical Writers*, 2 vols (ed. K. Bennett), Oxford University Press (2015), 271–7. [*Aubrey*]
Lisa Jardine, *On a Grander Scale: The Outstanding Life of Sir Christopher Wren*, HarperCollins (2002). [*Jardine*]

Ward's dispute with Hobbes, and his work on universal languages, are covered in depth in

Douglas M. Jesseph, *Squaring the Circle: The War between Hobbes and Wallis*, University of Chicago Press (1999). [*Jesseph*]
M. M. Slaughter, *Universal Languages and Scientific Taxonomy in the Seventeenth Century*, Cambridge University Press (1982). [*Slaughter*]
James Dougal Fleming, *The Mirror of Information in Early Modern England: John Wilkins and the Universal Character*, Palgrave Macmillan (2017).

Wren and his scientific work are covered in depth by

J. A. Bennett, *The Mathematical Science of Christopher Wren*, Cambridge University Press (1982). [*Bennett*]
D. Graham Burnett, *Descartes and the Hyperbolic Quest: Lens Making Machines and Their Significance in the Seventeenth Century*, American Philosophical Society (2005).

Further useful information can also be found in the multiple volumes of the following:

Philip Beeley and Christoph J. Scriba (eds), *The Correspondence of John Wallis*, Oxford University Press (2003–). [*Wallis*]
Eric G. Forbes, Lesley Murdin, and Frances Willmoth (eds), *Correspondence of John Flamsteed, First Astronomer Royal*, 3 vols, Institute of Physics (1995–2002). [*Flamsteed*]

Notes and references

1. J. and J. A. Venn, *Alumni Cantabrigienses: A Biographical List of All Known Students, Graduates and Holders of Office at the University of Cambridge, from the Earliest Times to 1900*, part 1, Vol. 4, Cambridge University Press (1927), 334–5.
2. The two main contemporary sources for Ward's life are *Pope* and *Aubrey*; both of them knew Ward well. Anthony à Wood's treatment of Ward in his *Athenæ Oxonienses, an Exact History of All the Writers and Bishops Who Have Had Their Education in the University of Oxford* (ed. P. Bliss), 3rd edn, Vol. 4 (London, 1820), cols 246–52, relies heavily on Aubrey's notes but has additional information.
3. For John Wallis, see the chapter by Philip Beeley in *Oxford Figures*, 28–53. The best source for Jeremiah Horrocks is Wilbur Applebaum's 'Horrocks [Horrox], Jeremiah' in the *ODNB* (2004).
4. See *Aubrey*, 125, 272, and J. F. Scott, 'Oughtred, William' in the *Dictionary of Scientific Biography*, Vol. 10, Charles Scribner's Sons (1974), 254. For the Oughtred quotation, see William Oughtred, *The Key of Mathematics New Forged and Filed*, London (1647), ix–x.
5. Ward was one of two candidates, and the vote was 52 to 50 in his favour. See J. E. B. Mayor, 'Seth Ward', *Notes and Queries* (Second Series) 7 (April 1859), 270.
6. See *Pope*, pp. 19–22, and *Aubrey*, p. 27.
7. See 'Trevor, Sir John' in the *ODNB* (2006).
8. Seth Ward, *De cometis* (Oxford, 1653), pp. 31–3. The claim of novelty was made in Robert Plot, *The Natural History of Oxford-shire, Being an Essay toward the Natural History of England* (Oxford, 1677), 230, but elliptical cometary orbits had been suggested earlier by Tycho Brahe, and also by Sir William Lower, in a letter to Thomas Harriot dated 6 February 1610/11; see Donald K. Yeomans, *Comets: A Chronological History of Observation, Science, Myth, and Folklore*, John Wiley & Sons (1991), 63–4. Although correct, Ward's suggestion seems not to have influenced later developments.
9. See Ian Roy and Dietrich Reinhart, 'Oxford and the Civil Wars' and Blair Worden, 'Cromwellian Oxford' in *Tyacke*, pp. 710–11 and 733–72.
10. Although Ward left no record on this, his contemporary biographer Anthony à Wood and Pope's modern editor J. B. Bamborough agree that he in fact did subscribe; see Bamborough, introduction to *Pope*, xxxii.
11. John Wallis, *A Defence of the Royal Society* (1678), quoted in J. F. Scott, *The Mathematical Work of John Wallis, D.D., F.R.S. (1616–1703)*, 2nd edn, Chelsea Publishing (1981), 10. This is substantially the same as what Wallis wrote in his MS autobiography: Christof J. Scriba, 'The Autobiography of John Wallis, F.R.S.', *Notes and Records of the Royal Society* 25 (1) (June 1970), 40.
12. See *Feingold*, and M. Feingold, 'The mathematical sciences and new philosophies', in *Tyacke*, pp. 359–448.
13. Richard Waller, 'The life of Dr. Robert Hooke', in R. Waller (ed.), *The Posthumous Works of Robert Hooke* (London, 1705), reprinted by Frank Cass & Co. (1971), viii: 'By the persuasion of Dr. Seth Ward, afterward Bishop of Salisbury, about 1656, he apply'd himself more particularly to the Study of Astronomy, and about 58 or 59, he says thus, "I have contriv'd several astronomical Instruments for making Observations, both at Sea and Land, which I afterwards produc'd before the Royal Society".

14. *Pope*, 33–4. Wood repeated the story, as did *Aubrey* on p. 300, and noted Henry Stubbe retailing the story to Thomas Hobbes, an enemy of both Ward and Wallis, in Thomas Hobbes, *The Correspondence of Thomas Hobbes*, Vol. 1 (ed. N. Malcolm), Oxford University Press (1994), 338–9 (letter from Stubbe to Hobbes, 9 November 1656).
15. This is from the dedication to Goddard of S. Ward, *In Ismaelis Bullialdi Astronomiae Philolaicae Fundamenta, Inquisitio Brevis* (1653). Ward was evidently unaware of Thomas Harriot's use of an astronomical telescope in 1609. See John W. Shirley, *Thomas Harriot: A Biography*, Oxford University Press (1983), 397–8.
16. John Ward, *The Lives of the Professors of Gresham College* (London, 1740), 90.
17. R. T. Gunther, 'The first observatory instruments of the Savilian Professors at Oxford', *The Observatory*, no. 758 (1937), 191.
18. See *Davies*, p. 321.
19. Letter from Ward to Sir Justinian Isham, 27 February 1651/2, quoted in H. W. Robinson, 'An unpublished letter of Dr Seth Ward relating to the early meetings of the Oxford Philosophical Society', *Notes and Records of the Royal Society of London* 7 (1) (December 1949), 68–70.
20. See the *ODNB* (2008), 'Reeve, Richard'.
21. Noel Malcolm and Jacqueline Stedall, *John Pell (1611–1685) and his Correspondence with Sir Charles Cavendish: The Mental World of an Early Modern Mathematician*, Oxford University Press (2005), 335–40.
22. A. D. C. Simpson, 'Robert Hooke and practical optics: Technical support at a scientific frontier', in Michael Hunter and Simon Schaffer (eds), *Robert Hooke: New Studies*, Boydell (1989), 37, n. 13. Wallis's letter, cited by Simpson, appears in *Wallis*, Vol. 1, 180, letter to Huygens dated 7 April 1656. See also Hooke's mention of the improvement of telescopes 'for Length and Goodness' by Neile, Wren, Goddard, and Reeve, in William Derham (ed.), *Philosophical Experiments and Observations of the Late Eminent Robert Hooke* (London, 1726), 390.
23. Letter from Ward to Oughtred, 4 January 1652/53, printed in Stephen Jordan Rigaud, *Correspondence of Scientific Men of the Seventeenth Century*, Vol. 1 (Oxford, 1841), reprinted by Georg Olms (1965), 75.
24. For a complete history of attempts to address the problem, see Peter Colwell, *Solving Kepler's Equation over Three Centuries*, Willmann-Bell (1993); for the solutions of Boulliau, Ward, and Horrocks, see pp. 9–11.
25. This and the following account of the Boulliau–Ward dispute summarize the analysis of Curtis Wilson, 'From Kepler's laws, so-called, to universal gravitation: Empirical factors', *Archive for History of Exact Sciences* 6 (1970), 89–170, especially pp. 106–22.
26. Boulliau and Ward also corresponded privately. See Robert A. Hatch, *The Collection Boulliau (BN, FF. 13019–13059): An Inventory*, American Philosophical Society (1982), 230, 231, 381: letters of January and July 1657 and March 1659. There is also a letter from Boulliau to Wallis, May 1655.
27. John Newton, *Astronomia Britannica, Exhibiting The Doctrine of the Sphere, and Theory of the Planets Decimally by Trigonometry, and by Tables, Fitted for the Meridian of London, according to the Copernican Systeme As it is illustrated by Bullialdus, and the easie way of Calculation, lately published by Doctor WARD* (London, 1657), 'to the Reader', vii–viii.
28. Thomas Streete, *Astronomia Carolina, A New Theorie of Coelestial Motions* (London, 1661), 14, where he credits 'Count Pagan, Dr. Seth Ward and others'.

29. See also the reference to Ward's geometrical methods by Sir Jonas Moore (1617–79) in *Moore's Arithmetick in Two Books*, Part 2, *Contemplationes Geometricae in Two Treatises*, London (1660), 19. William Whiston, Newton's immediate successor as Lucasian professor of mathematics at Cambridge, still felt the need to explicate Ward's and Boulliau's methods in his lectures, only to knock them down again. See William Whiston, *Astronomical Lectures, Read in the Publick Schools in Cambridge*, 2nd edn (London, 1728), 307–27, where he described Ward's 'famous' hypothesis and said, 'This Blot we shall endeavour to wipe off from Astronomy, by the following Geometrical Method which we propose'. Newton, however, was better disposed toward it: see D. T. Whiteside (ed.), *The Mathematical Papers of Isaac Newton*, Vol VI: *1684–1691*, Cambridge University Press (1974), 172 n. 163. On p. 171 Whiteside describes Ward's theory as 'elegant'.
30. *Aubrey*, p. 110.
31. See *Oxford Figures*, p. 12.
32. For Bainbridge's acquisition of the MS, see [*New College*], pp. 23–4, and G. J. Toomer, *Eastern Wisedome and Learning: The Study of Arabic in Seventeenth-Century England*, Oxford University Press (1996), 249–50.
33. Our account relies on *Jesseph*, which can probably not be bettered as a history of Hobbes, Wallis, and Ward; see also *Oxford Figures* or *Savilian Geometry*.
34. Allen G. Debus. *Science and Education in the Seventeenth Century: The Webster Ward Debate*, American Elsevier (1970); this book reprints in facsimile all the related texts.
35. Jan Prins, 'Ward's polemic with Hobbes on the sources of his optical theories', *Revue d'histoire des sciences* 46 (nos. 2/3) (1993), 195–224.
36. *Jesseph*, pp. 67–8.
37. *Pope*, pp. 125–6.
38. Robinson (note 19), pp. 69–70.
39. I rely almost exclusively upon *Slaughter*, who distinguishes universal from philosophical languages on pp. 2–3.
40. *Slaughter*, p. 133.
41. John Wilkins, *An Essay towards a Real Character, and a Philosophical Language* (London, 1668), 'To the reader'.
42. Debus (note 34), p. 214.
43. *Slaughter*, pp. 178–83.
44. See *Pope*, pp. 47–8, and Wood (note 2), col. 248.
45. *Pope*, p. 32.
46. Wood (note 2), col. 249.
47. Thomas Birch, *The History of the Royal Society of London for Improving of Natural Knowledge, from its First Rise*, Vol. I (London, 1756), 3–4 (reprinted by Georg Olms, 1968).
48. H. W. Turnbull (ed.), *The Correspondence of Isaac Newton*, Vol. 1: *1661–1675*; Cambridge University Press (1959), 73; letter from Henry Oldenburg to Newton, dated 2 January 1671/2. See also note 29.
49. John Harrison, *The Library of Isaac Newton*, Cambridge University Press (1978), 260, and letter to Nathanael Hawes, Esq., 26 May 1695, in H. W. Turnbull (ed.), *The Correspondence of Isaac Newton*, Vol. 3: *1688–1694*, Cambridge University Press (1961), 364. This is where Newton referred to Oughtred as 'a Man whose judgment (if any man's) may be safely relied upon'.

50. Alan Cook, *Edmond Halley: Charting the Heavens and the Seas*, Oxford University Press (1998), 377, and letter dated 31 January 1679/80, published in *Flamsteed*, Vol. 1, pp. 726–32.
51. *Pope*, p. 192.
52. For biographical information, see *Jardine*, but for Wren's early life and education, see *Davies*. Also worth consulting is Adrian Tinniswood, *His Invention so Fertile: A Life of Christopher Wren*, Oxford University Press (2001).
53. Bennett, pp. 14–15.
54. Christopher Wren, *Parentalia, or, Memoirs of the Family of the Wrens, viz. of Mathew Bishop of Ely, Christopher Dean of Windsor, &c. but Chiefly of Sir Christopher Wren* (London, 1750), 183.
55. Wren (note 54), p. 182.
56. *Jardine*, 70; *Davies*, p. 313.
57. Wren (note 54), 198–9. See also the comment by *Bennett*, p. 126, note 12.
58. William Oughtred, *Clavis mathematicae* (Oxford, 1652), ix.
59. *Davies*, p. 313.
60. This hypothesis is put forth by *Davies*, p. 318.
61. John Evelyn, *The Diary of John Evelyn* (ed. E.S. de Beer), Oxford University Press (1959), 339 (11 July 1654). Evelyn was a very distant relation; see *Davies*, p. 302, and *Jardine*, p. 89.
62. *Jardine*, p. 89.
63. *Davies*, p. 320.
64. See Wren (note 54), pp. 200–6, for an expanded English draft of the speech.
65. Anthony Geraghty, *The Sheldonian Theatre: Architecture and Learning in Seventeenth-Century Oxford*, Yale University Press (2013), 38.
66. See *Davies*, pp. 319, 321. Because the Savilian statutes stipulated that the professors lecture every week during term, the absences implied by Wren's records would not have been permissible.
67. C. A. Ronan and Harold Hartley, 'Sir Paul Neile, F.R.S. (1613–1686)', *Notes and Records of the Royal Society of London* 15 (July 1960), 159–65.
68. Albert Van Helden, 'Christopher Wren's 'De Corpore Saturni', *Notes and Records of the Royal Society of London* 23 (2) (Dec. 1968), 213–29; see also *Bennett*, pp. 28–32.
69. For the invention and early use of the telescopic micrometer, as well as the Gascoigne–Oughtred connection, see David Sellers, *In Search of William Gascoigne: Seventeenth-Century Astronomer*, Springer (2012).
70. For this and the next three paragraphs, see *Bennett*, pp. 44–54.
71. See Whiteside (note 29), pp. 310–11, and [*Principia*], Proposition XXXI, pp. 636–9; note that Newton does not credit Wren by name. See also Wilson (note 25), pp. 127–8.
72. This and the previous two paragraphs are based on *Bennett*, pp. 55–65 and 77–86, and *Jardine*, pp. 176–80 and 209–11.
73. For Wren's transition from astronomy to architecture, see Geraghty (note 65), pp. 35–42.
74. Henry Wotton, *The Elements of Architecture* (London, 1624), 1.
75. For Hooke's architecture, see Howard Colvin, *A Biographical Dictionary of British Architects: 1600–1840*, 3rd edn, Yale University Press (1995), 506–10.
76. *Jardine*, pp. 239–47; *Bennett*, p.91.
77. Wren (note 54), p. 198.
78. *Jardine*, pp. 168–71.
79. Wren (note 54), p. 260.

80. Simon Bradley and Nikolaus Pevsner, *Cambridgeshire*, The Buildings of England, 2nd edn, Yale University Press (2014), 165–7.
81. Wren (note 54), p. 260; the letter was published undated, except for the year.
82. For Bernard's life, see the excellent article by Hugh de Quehen in the *ODNB*. For a critical account of Bernard as a scholar, see Toomer (note 32), pp. 299–305.
83. Henry Oldenburg, *The Correspondence of Henry Oldenburg*, Vol. 5: *1668–1669* (ed. A. R. and M. B. Hall), University of Wisconsin Press (1968), 233, 235, 342.
84. Toomer (note 32), p. 299.
85. Cook (note 50), pp. 334–6.
86. Toomer (note 32), p. 178.
87. Dmitri Levitin, *The Kingdom of Darkness: Bayle, Newton, and the Emancipation of the European Mind from Philosophy*, Cambridge University Press (2022), 520–1. See also *Wallis*, Vol. 4, p. 6.
88. Oldenburg (note 83), Vol. 8: *1671–1672* (1971), 523–4.
89. *Wallis*, Vol. 4, pp. 45–8. For Gregory's solution, see Wilson (note 25), pp. 128–30.
90. Rigaud (note 23), Vol. 1, pp. 158–60.
91. *ODNB*; Oldenburg (note 83), Vol. 9: *1672–1673* (1973), 592–3.
92. Oldenburg (note 83), Vol. 12: *1675–1676* (1986), 200–1.
93. *Flamsteed*, Vol. 3, pp. 219, 534.
94. *Flamsteed*, Vol. 1, pp. 809–11; translation from Latin by the editors.
95. For more on the quadrant, see *Flamsteed*, Vol. 1, pp. 819–21, 831–3, 858.
96. *Flamsteed*, Vol. 1, pp. 818–9. Whatever Bernard saw is not listed in Gary W. Kronk, *Cometography: A Catalogue of Comets*, Vol. 1: *Ancient–1799*, Cambridge University Press (1999).
97. *Flamsteed*, Vol. 2, p. 956.
98. *Flamsteed*, Vol. 2, pp. 593, 599–601.
99. See the entry for Edward Bernard in the *ODNB*.

CHAPTER 3

Further reading

For general background, see

Andrew Warwick, *Masters of Theory: Cambridge and the Rise of Mathematical Physics*, University of Chicago Press (2003), 2–3.

For Gregory's inaugural address (translated by P. D. Lawrence and A. G. Molland), see

'David Gregory's inaugural lecture at Oxford', *Notes and Records of the Royal Society of London* 25 (1970), 143–78. [*Inaugural*]

Notes and references

1. See C. M. Eagles, 'David Gregory and Newtonian science', *British Journal for the History of Science* 10 (1977), 216–25.
2. Royal Society Library, London, MS 210; Edinburgh University Library, MS Dc.4.35; and Christ Church, Oxford, MS 131. For Gregory's critical approach to the *Principia* in the *Notae*, see Niccolò Guicciardini, *Reading the Principia: The Debate on Newton's Mathematical*

Methods for Natural Philosophy from 1687 to 1736, Cambridge University Press (1999), 180.

3. *Inaugural*, on p. 165.
4. Raymond Flood and John Fauvel, 'John Wallis', *Oxford Figures*, pp. 115–39, on p. 137.
5. The quotations in this section come from *Inaugural*, pp. 146 and 166–9.
6. John Keill, *Introductio ad veram astronomiam*, Henry Clements (1718), xiii.
7. Samuel Pepys, *Private Correspondence and Miscellaneous Papers of Samuel Pepys, 1679–1703* (ed. J. R. Tanner), G. Bell (1926), ii, 91–4; David Gregory, *Mercurius Oxoniensis* (1707).
8. David Gregory, *Elements of Astronomy*, J. Nicholson (1715), ii.
9. Daniel Waterland, *Advice to a Young Student*, John Crownfield (1730), 27–8.
10. Letter from Johann Bernoulli to Abraham De Moivre, 11 July 1705, Universitätsbibliothek Basel, L Ia 664, Nr. 2.
11. Thomas Johnson, *Quaestiones philosophicae* (Cambridge, 1741), 128, 130, 133.
12. Gregory (note 8), 'The Author's Preface', pp. xii and iv.
13. Natasha Bailey, 'The fate of the soul in mid-eighteenth-century Oxford: An undergraduate's conjectures', *History of Universities* 36 (2023), 205–22, on pp. 214–15.
14. *Statuta antiqua universitatis Oxoniensis* (ed. S. Gibson), Oxford University Press (1931), 529. Quoted in *Inaugural*, p. 155.
15. See Chapter 2 and Will Poole, 'Edward Bernard (1638–1697) and New College', *New College Notes* 14/3 (2020), 1–7, on p. 1.
16. From as early as 1690, Gregory had been working on Apollonius's *Problems*; see D. Gregory, 'Apollonii de rectarum linearum sectione proportionali . . ., c 1700', University of Edinburgh Library, Coll-33/Folio B [37].
17. See G. J. Toomer, *Apollonius: Conics Books V to VII: The Arabic Translation of the Lost Greek Original in the Version of the Banū Mūsā*, Springer-Verlag (1990), xxv. For Gregory's study of Apollonius's *Problems* (*c*.1690), see the University of Edinburgh Library, Coll-33/Folio B [21].
18. See, for example, Aldrich's note to Gregory on scientific Greek, University of Edinburgh Library, Coll-33/Folio C [196].
19. For Keill's hopes to check the final draft of Gregory's edition, see his letter to Sloane, 1702/3, British Library, London, Sloane MS 4039, fol. 78r. See also 'An account of a manuscript' (*c*.1703), in which Gregory comments on various editions of Euclid, University of Edinburgh Library. Coll. 33/Folio E [062]. Gregory's alleged partial translation of Euclid's *Phaenomena* is in the Bodleian Library, Savile Collection MS 1, fol. 327.
20. For the labour that Bernard put into editing the writings of Euclid, see Philip Beeley, '"A designe Inchoate": Edward Bernard's planned edition of Euclid and its scholarly afterlife in late seventeenth-century Oxford', *Reading Mathematics in Early Modern Europe: Studies in the Production, Collection, and Use of Mathematical Books* (ed. P. Beeley, Y. Nasifoglu, and B. Wardhaugh), Routledge (2021), 192–229.
21. Benjamin Wardhaugh, *Encounters with Euclid: How an Ancient Greek Geometry Text Shaped the World*, Princeton University Press (2001), 87.
22. Beeley (note 20), pp. 212–13.
23. For the importance of collaboration in early 18th-century scholarship, see Natasha Bailey, 'Academic collaboration in the Early Enlightenment: Daniel Waterland (1683–1740) and his Cambridge tyros', *English Historical Review* 139 (issue 596) (February 2024), 126–54.
24. *Inaugural*, p. 166.

25. Gregory's booklist from *c.*1693, University of Edinburgh Library, Coll-33/Folio E [040].
26. Gregory's *Geometria practica in usum academicorum Edinburgensium* (1685), University of Edinburgh Library, GB 0237 David Gregory Dc.1.75 Folio B [6].
27. Letter from Gregory to Hans Sloane, 24 April 1697, British Library, Sloane MS 4036, fol. 305r. Gregory also obtained a booklist on agriculture from an unidentifiable hand (possibly Jacob Bobart) around 1698. See University of Edinburgh Library, Coll-33/Folio B [33].
28. *Statuta antiqua universitatis Oxoniensis* (ed. S. Gibson), Clarendon Press, Oxford (1931), 529; quoted in *Inaugural*, p. 155.
29. David Gregory, *David Gregory, Isaac Newton and Their Circle: Extracts from David Gregory's Memoranda, 1677–1708* (ed. W. G. Hiscock), Oxford University Press (1937), 20.
30. Anita Guerrini, 'Chemistry teaching at Oxford and Cambridge, circa 1700', *Alchemy and Chemistry in the 16th and 17th Centuries* (ed. P. Rattansi and A. Clericuzio), Springer (1994), 183–99, on p. 190.
31. University of Edinburgh Library, Coll. 33/Folio E, fol. 69.
32. University of Edinburgh Library, Coll. 33/Quarto A [45].
33. See Tom Jones, *George Berkeley: A Philosophical Life*, Princeton University Press (2021), 52.
34. Richard Westfall, *The Life of Isaac Newton*, Cambridge University Press (2015), 272.
35. Allan Chapman and Christopher Hollings, 'A century of astronomers: From Halley to Rigaud', *Savilian Geometry*, 54–91, on p. 79.
36. David Irving, *Lives of Scottish Writers*, Vol. 2 (Adam and Charles Black, 1850), on p. 281.
37. See Anthony Wood, *Athenae Oxonienses* (F. C. and J. Rivington, 1820), iv, 737.
38. A copy of Gregory's scheme is enclosed in a letter from Arthur Charlett to Samuel Pepys on 15 October 1700, in *Memoirs of Samuel Pepys, Esq* (Henry Colburn, 1825), ii, 239–42.
39. *Euclid's Elements of Geometry, from the Latin Translation of Commandine* (translated by S. Cunn), T. Woodward (London, 1723); the citation is from the extended title.
40. Letter from Caswell to Flamsteed, 8 May 1705, in Francis Baily, *An Account of the Revd. John Flamsteed, the First Astronomer-Royal* (London, 1835), p. 240.
41. Christopher Wordsworth, *Scholae Academicae: Some Account of the Studies at the English Universities in the Eighteenth Century*, Frank Cass (1968), on p. 71.
42. E. G. R. Taylor, *The Mathematical Practitioners of Tudor and Stuart England*, Cambridge University Press, for the Institute of Navigation (1954), 125.
43. 'The solutions of three chorographic problems by a member of the Philosophical Society of Oxford', *Philosophical Transactions* 15 (31 December 1685), 1231–6.
44. For the reference to Caswell's bodily infirmity, see Irving (note 36), p. 268, and https://archives.balliol.ox.ac.uk//Archives/stcrossmemorials.asp#gsc.tab=0.
45. A. H. T. Robb-Smith, 'The life and times of Dr Richard Frewin', *Early Modern Practitioners Working Papers* (ed. M. Pelling), 1–38, on pp. 7, 9: https://practitioners.exeter.ac.uk/wp-content/uploads/2014/11/Frewin.pdf.
46. Thomas Burnet, *Telluris theoria sacra* (London, 1681); English translation, *Sacred Theory of the Earth* (1684).
47. John Keill, *An Examination of Dr. Burnet's Theory of the Earth*, H. Clements (Oxford, 1698), 12, 30.
48. Waterland (note 9), p. 25, Bodleian Library, Oxford, MS Rawl. D. 40, fol. 7r. For a brief discussion, see Natasha Bailey, 'Sleek Divines and Beaux Esprits: The Universities and the

System of Knowledge in England, *c.*1690–1750', DPhil thesis, University of Oxford, 2022, p. 109.
49. Waterland (note 9), pp. 25–6.
50. For these notes, see Cambridge University Library, Add MS 9317.
51. See the *ODNB*, John Henry, 'Keill, John (1671–1721)'.
52. See Guicciardini (note 2), p. 187.
53. Dmitri Levitin, *Kingdom of Darkness: Bayle, Newton, and the Emancipation of the European Mind from Philosophy*, Cambridge University Press (2022), 637–44.
54. Translation from Levitin (note 53); see also Louis Ashman, 'David Gregory's and John Keill's Newtonian pedagogy: Oxford and the Scottish reception of Isaac Newton's natural philosophy', *History of Universities* 36 (2023), 155–83, on p. 172.
55. Letter from Johann Bernoulli to William Burnet, 3 August 1710, Universitätsbibliothek Basel, L Ia 654, fol. 7r.
56. Letter from William Burnet to Johann Bernoulli, 29 August 1710, Universitätsbibliothek Basel, L Ia 654, fol. 13.
57. John Keill, 'Epistola ad clarissimum virum Edmundum Halleium Geometriæ Professorem Savilianum, de legibus virium centripetarum', *Philosophical Transactions* 26 (1708), 174–88, on p. 175.
58. Niccolò Guicciardini, 'The Newton–Leibniz Calculus Controversy, 1708–1730', *The Oxford Handbook of Newton* (ed. E. Schliesser and C. Smeenk), Oxford University Press (2017), *DOI: 10.1093/oxfordhb/9780199930418.013.9.*
59. John B. Shank, *The Newton Wars and the Beginning of the French Enlightenment*, University of Chicago Press (2008), 183.
60. Guicciardini (note 58).
61. For a detailed discussion, see Mordechai Feingold, 'Newton, Leibniz, and Barrow too: An attempt at a reinterpretation', *Isis* 84 (1993), 310–38, on pp. 325–6.
62. Cambridge University Library, Add. MS 9597/2/18, fol. 69v, consulted via The Newton Project (https://www.newtonproject.ox.ac.uk/view/texts/diplomatic/NATP00280).
63. Wordsworth (note 40), pp. 75–6.
64. A copy of the 1718 edition of the last of these works, from the private library of Richard Watson, Fellow of Trinity College, Cambridge, contains detailed annotations; thanks are due to Karen Thomson for this information.
65. Ashman (note 54), p. 175.
66. Draft of Keill's astronomy lectures at Cambridge University Library, Add MS 9597/6/51, fol. 3r.
67. Keill's astronomy lectures (note 66), fols. 8v–9r. For this point, Keill cites Plato's *Epinomis*.
68. John Friesen, 'Christ Church Oxford, the ancients–moderns controversy, and the promotion of Newton in post-Revolutionary England', *History of Universities* 33 (2008), 33–66, on p. 53.
69. Kristine Haugen, *Richard Bentley: Poetry and Enlightenment*, Harvard University Press (2011), 52.
70. Keill's lectures (note 66), fol. 2v.
71. John Keill (note 6), pp. 77 and 84–5, with translations from Levitin (note 53), pp. 649–50. See also David B. Wilson, *Seeking Nature's Logic: Natural Philosophy in the Scottish Enlightenment*, Pennsylvania State University Press (2009), 53 and 46–8.
72. Levitin (note 53), pp. 649–51.
73. Guicciardini (note 58).

74. John Keill, 'Epistola ad . . . Gulielmum Cockburn . . . In qua Leges Attractionis aliaque Physices Principia traduntur', *Philosophical Transactions* 26 (1708), 97–110.
75. Anita Guerrini, 'James Keill, George Cheyne, and Newtonian physiology, 1690–1740', *Journal of the History of Biology* 18 (1985), 247–66, on p. 255.
76. Antonio Clericuzio, *Elements, Principles and Corpuscles: A Study of Atomism and Chemistry in the Seventeenth Century*, Kluwer (2000), 103–4.
77. A book of College exercises, anno 1722, British Library, London, Lansdowne MS 394, fols. 11r–12r.
78. For a poem about their marriage, see University of Nottingham, Pw V 143.
79. John Freind, *Chymical Lectures*, Philip Gwilliam (London, 1712), 4.
80. Nigel Aston, *Enlightened Oxford: The University and the Cultural and Political Life of Eighteenth-Century Britain and Beyond*, Oxford University Press (2023), 206.
81. [John Greene], *The Academic: Or a Disputation on the State of the University of Cambridge*, C. Say (1750), 23–4.

CHAPTER 4

Further reading

Comprehensive accounts of James Bradley's life and work are

Stephen Peter Rigaud, *Miscellaneous Works and Correspondence of the Rev James Bradley, D.D., F.R.S.* (Oxford, 1832). [*Rigaud*]

John Fisher, *The Life and Work of James Bradley: The New Foundations of 18th Century Astronomy*, Oxford University Press (2023). [*Fisher*]

There is also an overview of Bradley's work by M. Williams in the *Dictionary of Scientific Biography.*

For further background material, particularly on the astronomical context of 17th-century Oxford, see

Lesley Murdin, *Under Newton's Shadow: Astronomical Practices in the Seventeenth Century*, Adam Hilger (1985).

See also the chapter:

G. L'E. Turner, 'The physical sciences', *The History of the University of Oxford*, Vol. 5: *The Eighteenth Century* (ed. L. S. Sutherland and L. G. Mitchell), Clarendon Press, Oxford (1986), 659–82.

The edition of Newton's *Principia* for this chapter is

Isaac Newton, *Philosophiae Naturalis Principia Mathematica*, 3rd edn (London, 1726), transl. by I. Bernard Cohen and Anne Whitman, London (1999). [*Principia*]

Notes and references

1. From a communication dated 4 September 1721 from Martin Foulkes to Archbishop Wake.

2. Non-jurors were people who had given an oath to James II and who refused to relinquish it after the 'Glorious Revolution' of 1688–9. They were all deprived of public offices and were removed from jury service. Many had sympathies with the Jacobite cause, and most were Tories or High Church Anglicans.
3. For Hearne's assertion and Rigaud's defence, see *Rigaud*, p. ix, and *Fisher*, Chapter 3, on p. 100.
4. See *Principia*, Book III, Proposition 42.
5. The Longitude Act was passed in July 1714, towards the end of Queen Anne's reign.
6. Bradley was elected a Fellow of the Royal Society at a Council meeting on 6 November 1718, chaired by Isaac Newton.
7. The 'Honourable Company', or 'The Honourable, the United Company trading to the East Indies', is usually known as the East India Company. See *Fisher*, Chapter 2, on p. 52, where there is a full account of the massacre and its aftermath.
8. See Robert Grant, *History of Physical Astronomy from the Earliest Ages to the Middle of the Nineteenth Century* (London, 1852), 82.
9. Wanstead House was built from 1715 to 1722; the architect was Colen Campbell, author of *Vitruvius Britannicus, or the British Architect* (see also note 41).
10. James Pound's accounts book reveals repeated visits to Molyneux's houses in Richmond and Kew.
11. Samuel Molyneux, 'A description of an instrument set up in Kew, in Surrey, for investigating the annual parallax of the fixed stars, with an account of the observations made therewith', in Rigaud (note 3), pp. 109–10. See also *Fisher*, Chapter 4, on p. 129.
12. James Gregory, *Geometriae pars universalis* (Padua, 1668).
13. See M. E. W. Williams, 'Flamsteed's alleged measurement of annual parallax for the polestar', *Journal for the History of Astronomy* 10 (1979), 102–16.
14. The author thanks Ian Ridpath, editor of the *Antiquarian Astronomer*, for this information; see John Fisher, 'James Bradley's discovery of the aberration of light', *Antiquarian Astronomer* 16 (June 2022), 1.
15. John Conduitt, *Memorabilia*, Keynes MS 130.5, dated 13 May 1730, King's College Library, Cambridge; see also Microfilm, MS 659. Department of Manuscripts, Cambridge University Library.
16. See *Principia*, Book III, Proposition 21, Theorem 17.
17. See Conduitt (note 15).
18. Graham's instrument had a much more sophisticated design than Hooke's of 1669, being essentially an adapted transit instrument. The star was observed passing through the centre of the field of vision, with each observed passage adjusted by an accurate screw micrometer.
19. See Williams (note 13), and Jacques Cassini, 'Reflexions sur une lettre de M. Flamsteed à M. Wallis touchant la parallaxe annuelle de l'étoile', *Mémoires de l'Académie Royale des Sciences pour 1699* (Paris, 1702), 177–83.
20. Dan Bogart, 'Turnpike trusts and the transportation revolution in 18th century England', *Explorations in Economic History* (2005), 19–30.
21. See John Fisher, Data extracted from James Bradley's Wanstead Observation Book. This is included in James Bradley, *Observations made on the fixed stars made at Wanstead from 1727 to 1747*, pp. 201–86, Bradley MS 15, Department of Western Manuscripts, Bodleian Library, Oxford.

22. James Bradley, 'A letter to Dr Edmond Halley, Astronomer Royal and etc. giving an account of a new discovered motion of the fixed stars', *Philosophical Transactions* XXXV (1729), 637–61.
23. W. H. McCrea, 'The significance of the discovery of aberration', *Quarterly Journal of the Royal Astronomical Society* (March 1963), 41–3.
24. See McCrea (note 23).
25. Eustachio Manfredi, *De annuis inerrantium stellarum aberrationibus* (On the Annual Aberrations of the Fixed Stars) (Bologna, 1729).
26. The motion of the Earth as a heresy was a consequence of the deliberations of the Council of Trent that met from 1545 to 1563. Copernicus had dedicated his *De revolutionibus* of 1543 to Pope Paul II, the very pope who called the Council.
27. Bradley's acting executor Samuel Peach withheld or destroyed all of Bradley's private personal archive.
28. Joseph Stedman Jones, 'Moses Principia or Principia Mathematica: Hutchinsonianism and Natural Philosophy in Mid-Eighteenth Century Oxford', MSc dissertation, University of Oxford, 2017.
29. Jane Austen's novels deal with contentious topics touching on the position of women in society and slavery.
30. The Rt Revd Benjamin Hoadly, commonly called the most controversial churchman of his age, believed that all should be allowed to interpret scripture for themselves. The American revolutionary Samuel Adams later cited Hoadly as an inspiration for American liberties and the Constitution of the United States of America. Hoadly remained a close friend of Bradley throughout his life.
31. Robert Lowth, Oxford's professor of poetry, later became Bishop of London.
32. See Jones (note 28).
33. The Radcliffe Observatory was begun in 1772, and by the following year it already functioned as a working observatory.
34. The astronomer George Parker FRS was 2nd Earl of Macclesfield from 1732 until his death in 1764, and President of the Royal Society from 1752.
35. See *Principia*, Book III, Proposition 21, Theorem 17. Newton wrote: 'The equinoctial points regress, and the world's axis, by a nutation in every annual revolution, inclines twice towards the ecliptic and twice returns to its former position'. This is clear by Book I, Proposition 66, Corollary 20; this motion of nutation, however, must be very small – 'either scarcely or not at all perceptible'.
36. The precession of the equinoxes goes through a complete cycle in a period of over 25,000 years.
37. For over eighteen years, the nutation resulted in a deviation of about 9 arc-seconds north and south. The lunar-based nutation of the Earth's axis also accelerated and decelerated the precession of the equinoxes by a similar amount.
38. Pierre Louis Moreau de Maupertuis lived from 1698 to 1759.
39. This refusal to accept obvious observational evidence is an example of *conceptual bias*, a refusal to accept evidence because it contradicts a well-established theory.
40. James Bradley was seeking horological corroboration for his hypothesis, claiming that he was observing a lunar-induced nutation due to the oblateness of the Earth.
41. Although they are sometimes confused, Colin Campbell FRS is not the same person as Colen Campbell, the architect of Wanstead House, who died in 1729. Colin Campbell, the owner

of the Jamaican estate, was undertaking the experiment during the 1730s. Both men were related to the dukes of Argyll.

42. Voltaire's *Lettres* were published in Great Britain, Ireland, and the American colonies as *Letters from England*.
43. See *Fisher*, Chapter 9.
44. James Bradley, Bodleian Library, Bradley MS 31: *James Bradley's Observations at Greenwich with Two Transit Instruments from 25th July 1742 to January 1742/43. Lord Macclesfield's Similar Observations at Shirburne, September to December 1742 and March 1743*, 44 folios; written in 1742–3.
45. James Bradley, *Observations: Transits of Major Stars including adjustments to the telescope and observational method*, written in 1743, Cambridge University Library, RGO3 MS 1.
46. James Bradley, *Observations using a Quadrant, of Major Stars, including adjustments to the telescope and observational method*, written in 1743–4, Cambridge University Library, RGO3 MS 8.
47. Letter from Nicolas-Louis de Lacaille to James Bradley (author's translation from the French), dated 28 July 1744.
48. Francis Baily, *An Account of the Rev. John Flamsteed, the First Astronomer Royal, Compiled from his Manuscripts and Other Authentic Documents, Never before Published* (London, 1835).
49. Bradley's second assistant (Charles Mason) reduced a catalogue of 387 stars that was published in the *Nautical Almanac* of 1773; this was republished in 1797 in Bradley's observations, Vol. 1 (ed. T. Hornsby).
50. In a letter dated 3 April 1861 to Bartholomew Price of Pembroke College, Oxford, George Biddell Airy asserted that Samuel Peach was no gentleman; in truth, Peach's conduct left Airy speechless.
51. Friedrich Wilhelm Bessel, *Fundamenta astronomæ* (Königsberg, 1818).
52. Simon Schaffer, 'Astronomers mark time: Discipline and the personal equation', *Science in Context* 2(1) (1988), 115–45.

CHAPTER 5

Further reading

For an overview of this period in Oxford, see

John Fauvel, 'Georgian Oxford', *Oxford Figures*, Chapter 8, pp. 180–201. [*Fauvel*]
Nigel Aston, *Enlightened Oxford*, Oxford University Press (2023). [*Aston*]
L. W. B. Brockliss, *The University of Oxford: A History*, Oxford University Press (2016). [*Brockliss*]

For more information about Thomas Hornsby, see

Jim Bennett, 'Thomas Hornsby', *Sedleian*, pp. 73–91. [*Bennett*]
Allan Chapman, 'Thomas Hornsby and the Radcliffe Observatory', *Oxford Figures*, Chapter 9, pp. 202–20. [*Chapman*]

For further details of the Radcliffe Observatory, see

Ivor Guest, *Dr John Radcliffe and his Trust*, London (1991). [*Guest*]

A. D. Thackeray, *The Radcliffe Observatory, 1772–1972*, Oxford University Press (1972). [*Thackeray*]

Notes and references

1. The most significant of these enterprises were James Bradley, *Astronomical Observations, made at the Royal Observatory at Greenwich, from the Year MDCCL to the year MDCCLXII...* (ed. T. Hornsby and A. Robertson), 2 vols (Oxford, 1798–1805), and James Bradley, *Miscellaneous Works and Correspondence...* (ed. S. P. Rigaud) (Oxford, 1832).
2. R. T. Gunther, *Early Science in Oxford*, Vol. XI: Oxford Colleges and Their Men of Science, Oxford University Press (1937), 371; Ruth Wallis, 'Cross-currents in astronomy and navigation: Thomas Hornsby FRS (1733–1810)', *Annals of Science* 57 (2010), 219–40, on pp. 220–2. For his New College Lane residence, see H. E. Bell, 'The Savilian professors' houses and Halley's observatory at Oxford', *Notes & Records of the Royal Society of London* 16 (1961), 179–86.
3. See Wallis (note 2), pp. 225–6, and *Bennett*.
4. T. Hornsby, 'A discourse on the parallax of the sun', and 'Observations on the eclipse of the sun, April 1, 1764, in a letter to the Right Honourable James, earl of Morton, Pres. R.S', *Philosophical Transactions* 53 (1763), 467–95, on pp. 476, 491, 495, and 54 (1764), 145–9.
5. T. Hornsby, 'On the transit of Venus in 1769', and 'An account of the observations of the transit of Venus and of the eclipse of the sun, made at Shirburn Castle and at Oxford', *Philosophical Transactions* 55 (1765), 32–64, and 59 (1769), 172–82.
6. T. Hornsby, 'An account of the observations' (note 5), p. 181. See also Aston, pp. 575, 640, especially n.189.
7. T. Hornsby, 'The quantity of the sun's parallax as deduced from the observations of the transit of Venus, on June 3, 1769', *Philosophical Transactions* 61 (1771), 574–9. For Hornsby and Herschel, see *Bennett*, pp. 84–5.
8. T. Hornsby, 'An inquiry into the quantity and direction of the proper motion of Arcturus; with some remarks on the diminution of the obliquity of the ecliptic', *Philosophical Transactions* 63 (1773), 93–125, on pp. 96–8, 105.
9. For proper motion in Right Ascension and Declination see Hornsby (note 8), pp. 109–10, 122–5. The current value for the decline of the obliquity of the ecliptic is 47.5 arc-seconds per century.
10. The Radcliffe Trust was created to oversee the distribution of funds from the will of the physician John Radcliffe; see *Guest*, pp. 227–9, and *Thackeray*.
11. *Guest*, pp. 230–4, and G. Tyack, 'The making of the Radcliffe Observatory', *Georgian Group Journal* 10 (2000), 122–40.
12. *Guest*, pp. 235–6, and J. Fisher, *The Life and Work of James Bradley*, Oxford University Press (2023), 285–90. For Bird's precision graduation techniques, which took into account the effects that temperature had on the expansion of the brass, see Allan Chapman, *Dividing the Circle: The Development of Critical Angular Measurement in Astronomy 1500–1800*, Ellis Horwood (1990), 71–7. In 1790 Hornsby drew up a list of his own instruments, which were appraised by the instrument-maker Edward Nairne and subsequently purchased for the University; see Jim Bennett, 'A lost cabinet of experimental philosophy', in *Cabinets of Experimental Philosophy in Eighteenth-Century Europe* (ed. J. Bennett and S. Talas), Brill (2013), 69–78.

13. *Guest*, pp. 235–9.
14. *Guest*, pp. 237–9.
15. *Guest*, pp. 241–6, and Tyack (note 11), pp. 129, 133–4.
16. Thomas Bugge, *An Observer of Observatories. The Journal of Thomas Bugge's Tour of Germany, Holland and England in 1777* (ed. K. M. Pedersen and P. de Clercq), Aarhus (2010), 105–6, 119, and 128. For the collection of rainwater data at the Observatory, which has provided a virtually continuous record of rainfall from 1772 to the present day, see Chapter 2 of S. Burt and T. Burt, *Oxford Weather and Climate since 1767*, Oxford University Press (2019).
17. Bugge (note 16), pp. 108–11, 118, 128.
18. Bugge (note 16), p. 123.
19. For references to Herschel in 1781 and 1789, see Ms. Radcliffe Trust d.17, fols 23^{r}–24^{v} and $20b^{v}$.
20. Henry Best, *Personal and Literary Memorials* (London, 1829), 219–21.
21. For Hornsby's working practices at the Observatory see *Chapman*, pp. 211–13.
22. See *Chapman*. For the quality of Hornsby's observations see A. Rambaut, 'A note on the unpublished observations made at the Radcliffe Observatory Oxford, between the years 1774 and 1838', *Monthly Notices of the Royal Astronomical Society* lx (January 1900), 256–93, on pp. 277–8, and H. Knox-Shaw, J. Jackson, and W. H. Robinson, *The Observations of the Reverend Thomas Hornsby . . . made with the Transit Instrument and Quadrant at the Radcliffe Observatory, Oxford, in the years 1774 to 1798*, Oxford University Press (1932), 9.
23. Bradley (note 1), Vol. 1, pp. i–xxviii; Wallis (note 2), pp. 227, 230–4; Eric Forbes, 'Bradley's astronomical observations', *Quarterly Journal of the Royal Astronomical Society* 6 (1965), 321–8, on pp. 322–3. The 2 March 1771 agreement between Bird and Hornsby for the Radcliffe Observatory instruments was signed in the presence of John Peach's brother, Samuel Peach, Jr.; the latter married Susannah Bradley in 1776.
24. See Bradley (note 1), Vol. 1, pp. i–ii and xxviii, Wallis (note 2), pp. 231–3, and Jon Topham, 'Science, mathematics, and medicine', *The History of Oxford University Press*, Vol. II: 1780 to 1896 (ed. S. Eliot), Oxford University Press (2013), 512–57.
25. Best (note 20), p. 221; George Valentine Cox, *Recollections of Oxford* (Oxford, 1868), 128–9. For the institutional response to the intellectual and political threats posed by radicalism see Aston, and Trevor Levere, 'Dr Thomas Beddoes at Oxford: radical politics in 1788–93 and the fate of the Regius Chair in Chemistry', *Ambix* 28 (1991), 61–9.
26. This biographical material is taken from W. Sedgwick and A. Yoshioka, 'Robertson, Abram (1751–1826), mathematician and astronomer', *ODNB* (4 October 2004), and from 'Obituary', *Gentleman's Magazine*, 1st ser., 97/1, Bradbury, Evans & Co (February 1827), 176–8, on p. 176.
27. See Cox (note 25), p. 146.
28. *Guest*, pp. 251–7.
29. *Gentleman's Magazine* (note 26); A. Robertson, *Sectionum conicarum libri septem accredit tractatus de sectionibus conicis et de scriptoribus qui eam doctrinam tradiderunt* (Oxford, 1792), and Joseph Torelli (ed.), *Archimedis quæ supersunt omnia cum Eutocii Ascalonitae commentariis* (Oxford, 1792). See also Topham (note 24), pp. 519–20.
30. Rigaud's account of the fate of the Harriot papers at Oxford and his exceptionally thorough treatment of Harriot's initial telescopic findings can be found in S. P. Rigaud, *Supplement to Dr. Bradley's Miscellaneous Works; with an Account of Harriot's Astronomical Papers* (Oxford, 1833). See also *Fauvel*, pp. 191–3.

31. F. W. Bessel (ed.), *Fundamenta Astronomiae pro anno MDCCLV: deducta ex observationibus viri incomparabilis James Bradley in specula astronomiae Grenovicensi per annos 1750–1762 institutis* (Königsberg, 1818); A. A. Rambaut, 'Note on the unpublished observations made at the Radcliffe Observatory, Oxford, between the years 1774 and 1838; with some results for the year 1774', *Monthly Notices of the Royal Astronomical Society* 60.4 (1900), 265–93, on p 265.
32. See the preface to the 1805 edition of James Bradley, *Astronomical Observations, made at the Royal Observatory at Greenwich, from the Year MDCCL to the year MDCCLXII. . .* (ed. T. Hornsby and A. Robertson), 2 vols (Oxford, 1798–1805).
33. *Guest*, p. 252, citing Bodleian Ms. Dd. Radcliffe, c.52. Minutes of Trustees' meeting, 7 June 1811.
34. *Fauvel*, pp. 196–7, and see the George Chinnery papers, Christ Church, Oxford.
35. 'Notebooks of prof. Rigaud: Lectures on Astronomy delivered in about 1813 at Oxford, probably by Abram Robertson and in his hand', Bodleian Library Special Collections, Savile Collection, MS Rigaud 55, fols 2r and 5r.
36. Notebooks (note 35), fols 12–14, 18, 20.
37. For biographical material on Stephen Peter Rigaud, see 'Professor Rigaud [obituary]', *Gentleman's Magazine Review*, E. Cave (May 1839), 542–3, and John Rigaud, *Stephen Peter Rigaud, M.A., F.R.S., Formerly Savilian Professor of Astronomy and Radcliffe Observer, Oxford: A Memoir* (Oxford, 1883).
38. J. Rigaud (note 37), p. 5, and *Gentleman's Magazine* (note 37), p. 543.
39. See *Fauvel*, pp. 194–6, and *Brockliss*, pp. 235–9.
40. Notebooks (note 35), 'Similar scientific collections', Bodleian Library Special Collections, Savile Collection, MS Rigaud 49, and Letters from Alex Galloway to Stephen Peter Rigaud, 3, 19, and 23 November 1814; see also 'Letters written to prof. S.P. Rigaud', F-M, in Bodleian Library Special Collections, MS Rigaud 61.
41. Lists of undergraduates who attended prof. Rigaud's Savilian lectures, Bodleian Library Special Collections, MS. Savile e.10.
42. See Manuel Johnson, *Astronomical Observations Made at the Radcliffe Observatory for the Year 1840*, Vol. 1 (Oxford, 1842), xiii–xxii; R. T. Gunther, *Early Science in Oxford*, Vol. II: Astronomy, Oxford University Press (1923), 326.
43. For Baden Powell, see P. Corsi, *Science and Religion: Baden Powell and the Anglican Debate, 1800–1860*, Oxford University Press (1988), and Keith Hannabuss, 'The 19th century', *Oxford Figures*, Chapter 10, pp. 222–37.
44. Bradley (note 1).
45. Rigaud, (note 37) 'Memoir', 12 and 14 (citing the obituary in the *Edinburgh Review*). Rigaud's principal publications after his work on Bradley were his 1833 '*Supplement*' to the *Miscellaneous Works* (which contained his essay on Harriot's astronomical work); *Historical Essay on the first publication of Newton's Principia* (Oxford, 1838); *An Account of some early Proposals for Steam Navigation* (Oxford, 1838); and *Correspondence of Scientific Men of the Seventeenth Century: including letters of Barrow, Flamsteed, Wallis and Newton, printed from the originals in the collection of the Honorable the Earl of Macclesfield* (Oxford, 1841).
46. Rambaut (note 31), p. 266.

CHAPTER 6

Further reading

For the new astronomy and the personalities and disputes, see

J. L. E. Dreyer and H. H. Turner, *History of the Royal Astronomical Society*, Vol. 1: *1820–1920*, Royal Astronomical Society, London (1923), reprinted by Blackwell (1987). [*History*]

Ivor Guest, *Dr John Radcliffe and his Trust*, Radcliffe Trustees, London (1991) and online (2015): https://theradcliffetrust.org/wp-content/uploads/9780950248233_EPDF.pdf. [*Guest*]

For the development of Oxford's sciences and infrastructure, see

A. V. Simcock, *The Ashmolean Museum and Oxford Science 1683–1983*, History of Science Museum, Oxford University Press (1984). [*Ashmolean*]

For a careful near-contemporary appreciation of Pritchard, see

Ada Pritchard, *Charles Pritchard, D.D.; F.R.S.; F.R.A.S.; F.R.G.S.; Late Savilian Professor of Astronomy in the University of Oxford: Memoirs of His Life*, Seeley, London (1897) [*Pritchard*]; see in particular pp. 214–316 by H. H. Turner on 'Astronomical Work of Charles Pritchard'. [*Turner*]

Notes and references

1. M. G. Brock, *The History of the University of Oxford*, Vol. VII: Nineteenth-Century Oxford, Part 2 (ed. M. G. Brock and M. C. Curthoys), Clarendon Press, Oxford (2000), pp. 2, 9.
2. *Ashmolean*, pp. 11–13.
3. G. B. Airy, 'Report on the progress of astronomy during the present century', *Second Report of the British Association* (1932), 184.
4. *Guest* (2015), figures extracted from the Trust Accounts.
5. Devonshire Royal Commission on Scientific Instruction and the Advancement of Science, Third Report (1873), 43, paragraph 128.
6. Richard Brown, Fellow of the Royal Historical Society, https://richardjohnbr.blogspot.com/2011/02/university-education-1800-1870.html.
7. Sources include the insightful account in Mark Curthoys, 'Johnson, George Henry Sacheverell (1808–1881)', *ODNB* (2004 online).
8. Anonymous, 'Remarks on "An Address to the Members of Convocation"' (Oxford, 1837), attributed to G. H. S. Johnson and W. Baxter in the Bodleian library pre-1920 catalogue.
9. S. L. Ollard, 'The Oxford Architectural and Historical Society and the Oxford Movement', *Journal of the OAHS* (1939), 155; https://oxoniensia.org/volumes/1940/ollard.pdf.
10. 'Election and substitution, 25 April 1839' of G. H. S. Johnson, Bodleian Libraries, Oxford University Archive, OUA/Nep/A/1.
11. *Guest* (2015), on pp. 261–2.
12. Dr Amy Ebrey, Assistant Archivist of The Queen's College. has supplied details of George Johnson's income and clarified that he was not elected to a fellowship in 1829, but was a Taberdar.

13. See Curthoys (note 7).
14. Curthoys (note 7).
15. The Very Rev. G. Henry S. Johnson, 'The Vicarage of St Cuthbert's: The Enemies of the Dean', letter in *The Weston-super-Mare Gazette*, 17 September 1870.
16. G. H. S. Johnson obituary, *The Wells Journal* (10 November 1881), Wells, Somerset.
17. Robin Darwall-Smith, *A History of University College, Oxford*, Oxford University Press (2008), 354–5, with thanks for supplying details of Donkin's earnings from the college's Bursar's books and for offering detail on the honorary fellowship and memorial.
18. 'William Fishburn Donkin', *Monthly Notices of the RAS* 30 (1870), 85. This is unsigned, but the author is believed to be Bartholomew Price.
19. Keith Hannabuss, 'The mid-nineteenth century' and 'Henry Smith' in *Oxford Figures*.
20. I. Grattan-Guinness, 'Donkin, William Fishburn', *ODNB* (online, 2004).
21. See note 18.
22. A brief note of the controversy by H. F. Newall is in *History*, on pp. 158–9.
23. Hereford B. George, *New College 1856–1906*, Oxford University Press (1906), 17–19. George was a fellow there at age 18 in 1856 and a tutor from 1867 to 1891, and so lived through the changes.
24. Christopher Hollings, 'Four centuries of Sedleian professors', in *Sedleian*, pp. 19, 21, with thanks for finding the new statute, Hebdomadal Council Minutes 1854–66, OUA HC1/2/1, and the letter from A Member of Congregation, 'The new professorial statute' (May 1857), *University Notices October 1856–July 1859*, Bodleian Library, GA Oxon b.29.
25. The Radcliffe professorship – see Devonshire (note 5), p. 38 para 101.
26. Michael Stansfield of New College has gleaned these details from minutes of the Warden and thirteen discussions in 1856 (NCA 9793 at f.174r and f.181), published Ordinances of 1857 corrected 1870 (pp. 21–2) on annual payments, and has located other related New College documents.
27. Sir Roundell Palmer, 'Opinion 18 March 1858 for New College', New College Archive NCA 671, 4.
28. New College, Minutes of the Warden & Fellows, 'Ordinary College meeting 23 Feb. 1870', NCA.MIN/W&F1, 120 for the invitation, and 'Annual Accounts of Income and Expenditure', first entry for Savilian payment 1871, NCA 844.
29. *Guest* (2015), on p. 275, citing: 'Suggestion of the course which the Radcliffe Trustees should follow in filling up the vacant appointment of the Observer at Oxford, 12 November 1859', Bodleian Library, Bodl. MSS. DD. Radcl., c.40.
30. Pritchard's obituary for Robert Main appears in *The Observatory* 14 (June 1878), 55–6.
31. The quotations here are from G. B. Airy to G. H. Richards at the Admiralty, letter 13 September 1872, RGO 6 150, 158; cited in *Observatories*, p. 151.
32. *Observatories*, pp. 152–3, 223–4.
33. H. P. Hollis, in *History*, pp. 173–8.
34. Devonshire (note 5), on p. 43, paragraph 128.
35. Devonshire (note 5), on pp. 54–8.
36. John Fauvel, 'James Joseph Sylvester', *Oxford Figures*, Chapter 13, pp. 280–302.
37. Anthony J. Croft, *Oxford's Clarendon Laboratory*, Department of Physics, University of Oxford (1986), Part 3, pp. 27–35; the quotations are from pp. 31–2 and the full text is online at https://www.physics.ox.ac.uk/about-us/our-history.
38. See *Pritchard*.

39. A. M. Clerke, revised by Anita McConnell, 'Pritchard, Charles (1808–1893)', *ODNB* (online 2006).
40. John Toman, 'Francis Kilvert and Charles Pritchard – Clapham connections', *Clapham Society Local History Series* 8 (online, The Clapham Society, 2011).
41. *Pritchard*, on p. 71.
42. H. F. Newall, *History*, pp. 164–5.
43. William E. Plummer, 'Rev. Charles Pritchard D.D., F.R.S.', *The Observatory* 16 (1893), 258.
44. *Turner*, on p. 261.
45. *Ashmolean*, 'The growth of the Science area', pp. 14–17, on p. 15.
46. The fullest account of Phillips's role is in R. Hutchins, 'John Phillips, "Geologist–astronomer" and the origins of the Oxford University Observatory, 1853–1875', *History of Universities* 13 (1994), 194–249; Pritchard's quote is from his account to the Royal Astronomical Society's meeting on 14 December 1873, *The Astronomical Register* 12 (January 1874), 4–11. For a summary, see *Observatories*, pp. 42–51.
47. *Oxford University Gazette*, 111 (February 1873), 42–3.
48. Pritchard, *Astronomical Register*, on p. 6.
49. *Turner*, on pp. 276–7.
50. *New Statutes made for the University and Colleges of Oxford 1882, under the Universities of Oxford and Cambridge Act 1877* (1882), University 34 (1) and (4), New College 7 (1), (3), (19). Bodleian Library.
51. *Turner*, on pp. 269, 274.
52. *Turner*, on pp. 309–10.
53. *Turner*, on pp. 312–13.
54. *Pritchard*, on pp. 142–5.
55. *Turner*, on p. 316.
56. *Observatories*, Table 7.3 on pp. 378–9.

CHAPTER 7

Further reading

Useful biographies are:

H. C. [Plummer], 'Professor H. H. Turner', obituary, *Monthly Notices of the Royal Astronomical Society*, 91 (1931), 321–34. [*Plummer*]

R. A. [Sampson], 'Herbert Hall Turner – 1861–1930', *Proceedings of the Royal Society A* 133 (1931), i–ix. [*Sampson*]

W. H. McCrea, 'Harry Hemley Plaskett: 5 July 1893–26 January 1980', *Biographical Memoirs of Fellows of the Royal Society* 27 (1981), 444–78. [*McCrea*]

M. G. Adam, 'Harry Hemley Plaskett', *Quarterly Journal of the Royal Astronomical Society* 21 (1980), 486–8. [*Adam 1*]

Geoffrey Smith, 'Prof. Donald Blackwell 1921–2010', *Astronomy & Geophysics* 52 (2011), 3.37–3.38. [*Smith*]

An excellent and readable guide to the *Carte du Ciel* project is

H. H. Turner, *The Great Star Map*, John Murray (1912),

while a good recent account is

Ileana Chinnici, 'Precursors to IAU: Paris Observatory and the *Carte du Ciel* project', Thierry Montmerle and Danielle Fauque (eds), *Astronomers as Diplomats: When the IAU Builds Bridges Between Nations*, Cham: Springer (2022), 3–44.

An invaluable source for the history of the Oxford University Observatory in Plaskett's time is

M. G. Adam, 'The changing face of astronomy in Oxford (1920–1960)', *Quarterly Journal of the Royal Astronomical Society* 37 (1996), 153–79. [*Adam 2*]

A readable account of Arthur Milne's life and work, and of mathematics and the physical sciences at Oxford from the 1920s to around 1950, is

Meg Weston Smith, *Beating the Odds: The Life and Times of E. A. Milne*, Imperial College Press (2023).

Useful technical background about Plaskett and Blackwell's solar research can be found in

Karl Hufbauer, *Exploring the Sun: Solar Science since Galileo*, Johns Hopkins University Press (1991);

Jean-Louis Tassoul and Monique Tassoul, *A Concise History of Solar and Stellar Physics*, Princeton University Press (2004).

Finally, numerous detailed essays on the technical history of early 20th-century astronomy can be found in

Owen Gingerich (ed.), *The General History of Astronomy*, Vol. 4: *Astrophysics and Twentieth-century Astronomy to 1950: Part A*, Cambridge University Press (1984). [*Gingerich*]

The author is grateful to Professor Phil Charles, Dr Jon Godwin, and Dr Tony Lynas-Gray for sharing their memories of working with Donald Blackwell and explaining his scientific work.

Notes and references

1. Andrew Warwick, *Masters of Theory: Cambridge and the Rise of Mathematical Physics*, University of Chicago Press (2003), especially p. 240.
2. See *Plummer* and *Observatories*, pp. 197–8.
3. H. H. [Turner], 'Sir William Henry Mahoney Christie, K.C.B., F.R.S.', *The Observatory* 45 (1922), 77–81; see also *Observatories*, p. 198.
4. Letter from Christie to Secretary of Admiralty, 25 August 1884, Royal Greenwich Observatory Archives, Cambridge University Library, RGO 7/8.
5. See *Plummer* and *Sampson*.
6. See John Lankford, 'The impact of photography on astronomy', *Gingerich*, pp. 16–39, and Erich Robert Paul, *The Milky Way Galaxy and Statistical Cosmology, 1890–1924*, Cambridge University Press (1993), 83.
7. See *Plummer*.

8. Memorandum (no author) headed 'Candidates for Savilian Professorship of Astronomy 1893', 1 December 1893, Royal Greenwich Observatory Archives (see note 4), RGO 7/251.
9. See *Observatories*.
10. Printed letter from H. H. Turner, 'To the Electors of the Savilian Professor of Astronomy', October 1893, and offprint (undated) of a paper by Turner, 'On the reduction of measures of photographic plates', Royal Greenwich Observatory Archives (see note 4), RGO 7/251.
11. See *Sampson*.
12. See Lankford (note 6).
13. On the notion of the Royal Observatory as a 'factory', see Allan Chapman, 'Sir George Airy (1801–1892) and the concept of international standards in science, timekeeping and navigation', *Vistas in Astronomy* 28 (1985), 321–8, Simon Schaffer, 'Astronomers mark time: Discipline and the personal equation', *Science in Context* 2 (1988), 115–45, and Robert W. Smith, 'A national observatory transformed: Greenwich in the nineteenth century', *Journal for the History of Astronomy* 22 (1991), 5–20.
14. Letter from Turner to Christie, 14 January 1901. Royal Greenwich Observatory Archives (see note 4), RGO 7/241.
15. David Sellers, 'The gardener who became an astronomer: The life of Tom John Moore (*c.*1858–1924) of Hatfield, Yorkshire', *Society for the History of Astronomy Bulletin* 39 (2023), 13–18.
16. See *Observatories*, pp. 321–38.
17. See *Sampson*.
18. See *Plummer*.
19. See *Observatories*, pp. 325, 328–9.
20. See *Observatories*, pp. 330–6, and *Sampson*.
21. See *Observatories*, pp. 330–6.
22. These quotations from *The Observatory* are all from Matthew Stanley, '"An Expedition to Heal the Wounds of War": The 1919 eclipse and Eddington as Quaker adventurer', *Isis* 94 (2003), 57–89.
23. See *Sampson* and *Plummer*.
24. H. H. Turner, *The Great Star Map*, John Murray (1912).
25. Graeme J. N. Gooday and Morris F. Low, 'Technology transfer and cultural exchange: Western scientists and engineers encounter late Tokugawa and Meiji Japan', *Osiris* 13 (1998), 99–128; letter from Turner to Christie, 22 October 1895, Royal Greenwich Observatory Archives (note 4), RGO 7/196 (enclosing a letter from John Milne); A. L. Herbert-Gustar and P. A. Nott, *John Milne: Father of Modern Seismology*, Paul Norbury Publications (1980).
26. See *Observatories*, pp. 339–48.
27. See *Observatories*, pp. 341–8, Brian Kennett, 'Deep earthquake study still resonant at 100', *Astronomy & Geophysics* 63 (2022), 5.34–5.37, and Roger Hutchins, 'Bellamy, Frank Arthur (1863–1936)', *ODNB* (2004).
28. See *Observatories*, p. 365, and Roger Hutchins, 'Turner, Herbert Hall', *ODNB* (2004).
29. *Radcliffe*, pp. 90–1.
30. See *Observatories*, p. 336.
31. See *Observatories*, pp. 348–62.
32. See *McCrea* and Peter R. Broughton, *Northern Star: J. S. Plaskett*, University of Toronto Press, 2018.

33. See *McCrea* and Margaret Wilson, *Ninth Astronomer Royal: The Life of Frank Watson Dyson*, W. Heffer & Sons (1951), 185.
34. Meg Weston Smith, *Beating the Odds: The Life and Times of E. A. Milne*, Imperial College Press (2013), 136–7.
35. See Weston Smith (note 34), pp. 136–7, and *Observatories*, pp. 386–7.
36. See *Observatories*, pp. 387–9, and Weston Smith (note 34), pp. 137–8.
37. See *McCrea* and *Adam 1*.
38. See *Observatories*, pp. 92–3.
39. See *McCrea*, *Adam 2*, and Lee T. Macdonald, *Kew Observatory and the Evolution of Victorian Science 1840–1910*, University of Pittsburgh Press (2018).
40. See *Adam 2* (note 39).
41. See *McCrea*, *Adam 2*, and Weston Smith (note 34), p. 168.
42. See *Adam 2*, p. 167.
43. See *McCrea*, *Adam 2*, and R. J. Tayler (ed.), *History of the Royal Astronomical Society*, Vol. 2, Blackwell Scientific Publications (1987), 99–104.
44. H. H. Plaskett, 'Address delivered by the President, Professor H. H. Plaskett, on astronomical telescopes', *Monthly Notices of the Royal Astronomical Society* 106 (1946), 80–94.
45. See Lee T. Macdonald, 'The origins and construction of the Isaac Newton telescope, Herstmonceux, 1944–1967', *Journal of the British Astronomical Association* 120 (2010), 73–86, Lee T. Macdonald, '"A large chunk of glass": The 98-inch mirror of the Isaac Newton telescope, 1945–1959', *The Antiquarian Astronomer*, Issue 6 (2012), 4–18, and Anthony Wilson, *The Isaac Newton Telescope at Herstmonceux and on La Palma*, Science Projects Publishing, Herstmonceux (2010).
46. See *Adam 2* and *McCrea*.
47. See *McCrea*.
48. See *Adam 2*.
49. See *Adam 2*, on p. 177.
50. See *Smith*, Mark Hurn, 'David W Dewhirst 1926–2012', *Astronomy & Geophysics* 54 (2013), 1.36, Karl Hufbauer, *Exploring the Sun: Solar Science since Galileo*, Johns Hopkins University Press (1991), 91, and 'Blackwell, Prof. Donald Eustace', *Who Was Who* (online edition).
51. See *Smith*, Hurn (note 50), and R. O. Redman, 'The JPEC', *Quarterly Journal of the Royal Astronomical Society* 12 (1971), 39–44.
52. See *Smith*.
53. D. E. Blackwell, 'Future of optical astronomy', *Nature* 195 (1962), 854–6; 'Oxford University Department of Astrophysics (University Observatory)', *Quarterly Journal of the Royal Astronomical Society* 5 (1964), 379–82; see also *Adam 2*, p. 177.
54. See Blackwell (note 53), p. 855.
55. 'Department of Astrophysics: University of Oxford', *Quarterly Journal of the Royal Astronomical Society* 12 (1971), 328–33; Phil Charles, personal communication with the author, 20 October 2023.
56. George F. R. Ellis and Roger Penrose, 'Dennis William Sciama: 18 November 1926–19 December 1999', *Biographical Memoirs of Fellows of the Royal Society* 56 (2010), 401–22.
57. See *Smith* and 'Department of Astrophysics: University of Oxford', *Quarterly Journal of the Royal Astronomical Society* 23 (1982), 388–98, and 30 (1989), 345–55, on p. 350; see also Phil Charles (note 55).
58. See *Smith*; Jon Godwin, personal communication with the author, 25 July 2023.

59. See *Smith*, Godwin (note 58); also an interview with Phil Charles, 2 October 2023, and Tony Lynas-Gray, personal communication with the author, 14 and 24 October 2023.
60. See *Smith*.
61. See Lynas-Gray (note 59) and Godwin (note 58).
62. Interview with Phil Charles (note 59).
63. Interview with Phil Charles (note 59).
64. See *Adam 2*, pp. 177–8.
65. See *Observatories*, p. 438.

CHAPTER 8

Further reading

A description of the current understanding of the cosmos can be found in

J. Dunkley, *Our Universe: An Astronomer's Guide*, Belknap Press (2019).

The rise of physical cosmology and its setting in the history of general relativity can be found in

P. G. Ferreira, *The Perfect Theory: A Century of Geniuses and the Battle over General Relativity*, Little, Brown (2014). [*Ferreira*]

An in-depth description of the development of modern cosmology and the paradigms of dark matter and dark energy is given in

Richard Panek, *The Four Percent Universe: Dark Matter, Dark Energy and the Race to Discover the Rest of Reality*, Oneworld (2012).

The author is grateful to George Efstathiou and Joe Silk for allowing him to interview them at length, so that he could attempt to reconstruct their lives and careers, and also to Phil Charles, Roger Davies, Lance Miller, and Pat Roche for their collective 'institutional memory'.

Notes and references

1. See *Ferreira*.
2. See G. Ellis, A. Lanza, and J. Miller, *The Renaissance of General Relativity and Cosmology*, Cambridge University Press (1993).
3. D. Layzer, 'A preface to cosmogony. I. The energy equation and the virial theorem for cosmic distributions', *Astrophysical Journal* 138 (1963), 174–84.
4. A. Penzias and R. Wilson, 'A measurement of excess antenna temperature at 4080 mc/s', *Astrophysical Journal* 142 (1965), 419–21.
5. J. Silk, 'Cosmic black body radiation and galaxy formation', *Astrophysical Journal* 151 (1968), 459–71.
6. P. J. Peebles and J. Yu, 'Primeval adiabatic perturbation in an expanding universe', *Astrophysical Journal* 162 (1970), 815–36.
7. M. Srednicki and J. Silk, 'Cosmic ray anti-protons as a probe of a photino dominated universe', *Physical Review Letters* 53 (1984), 624–7.
8. See S. Aarseth, J. R. Gott, and E. Turner, '*N*-body simulations of galaxy clustering I. Initial conditions and galaxy collapse times', *Astrophysical Journal* 228 (1979), 664–83.

9. G. Efstathiou, 'On the Rotation and Clustering of Galaxies', PhD thesis, University of Durham, 1979.
10. J. R. Bond, G. Efstathiou, and J. Silk, 'Massive neutrinos and the large-scale structure of the Universe', *Physical Review Letters* 45 (1980), 1980–4.
11. M. Davis, G. Efstathiou, C. Frenk, and S. White, 'The evolution of large-scale structure in a universe dominated by cold dark matter', *Astrophysical Journal* 292 (1985), 371–94.
12. M. Wilson and J. Silk, 'On the anisotropy of the cosmological background of matter and radiation distribution. I. The radiation anisotropy in a spatially flat universe', *Astrophysical Journal* 243 (1981), 14–25.
13. J. R. Bond and G. Efstathiou, 'The statistics of cosmic microwave background radiation fluctuations', *Monthly Notices of the Royal Astronomical Society* 226 (1987), 655–87.
14. S. J. Maddox, W. J. Sutherland, G. Efstathiou, and J. Loveday, 'The APM galaxy survey: I. APM measurements and star-galaxy separation', *Monthly Notices of the Royal Astronomical Society* 243 (1990), 692–712.
15. G. Efstathiou, W. J. Sutherland, and S. J. Maddox, 'The cosmological constant and cold dark matter', *Nature* 348 (1990), 705–7.
16. G. Smoot *et al.* (COBE collaboration), 'Structure in the COBE differential microwave radiometer first year maps', *Astrophysical Journal Letters* 396 (1992), L1–L5.
17. See https://www.esa.int/Enabling_Support/Operations/Planck.
18. See J. Peacock *et al.* (2dF collaboration), 'A measurement of the cosmological mass density from clustering in the 2dF galaxy redshift survey', *Nature* 410 (2001), 169–73.
19. See https://www.zooniverse.org/.
20. G. Bertone, D. Hooper, and J. Silk, 'Particle dark matter: Evidence, candidates and constraints', *Physics Reports* 405 (2005), 279–390.
21. P. Gondolo and J. Silk, 'Dark matter annihilation at the galactic center', *Physical Review Letters* 83 (1999), 1719–22.
22. M. Bañados, J. Silk, and S. West, 'Kerr black holes as particle accelerators to arbitrarily high energy', *Physical Review Letters* 103 (2009), 111102.
23. J. Silk and M. Rees, 'Quasars and galaxy formation', *Astronomy and Astrophysics* 331 (1998), L1–L4.
24. K. Chamcham, J. Silk, J. D. Barrow, and S. Saunders, *The Philosophy of Cosmology*, Cambridge University Press (2017).
25. See https://www.kavlifoundation.org/.
26. P. A. R. Ade *et al.* (Planck collaboration), '*Planck* 2013 results. XVI. Cosmological parameters', *Astronomy & Astrophysics* 571 A16 (November 2014), 1–69.
27. G. Efstathiou, 'A lockdown perspective on the Hubble tension (with comments from the SHOES team)', arXiv:2007.10716 (2020).
28. Amon and G. Efstathiou, 'A non-linear solution to the S_8 tension?', *Monthly Notices of the Royal Astronomical Society* 516 (4) (2022), 5355–66.
29. See J. Silk, *Back to the Moon: The next giant leap for humankind, Princeton University Press* (2022).

CHAPTER 9

NOTES AND REFERENCES

1. S. Balbus and K. Brecher, Tidal friction in the binary pulsar system PSR 1913+16, *Astrophysical Journal* 203(1) (1976), 202–5.

2. G. B. Field, D. W. Goldsmith, and H. J. Habing, Cosmic-ray heating of the interstellar gas, *Astrophysical Journal Letters* 155 (1969), L149.
3. C. F. McKee and J. P. Ostriker, A theory of the interstellar medium – Three components regulated by supernova explosions in an inhomogeneous substrate, *Astrophysical Journal* 218 (1977), 148–69.
4. Steven A. Balbus and John F. Hawley, A powerful local shear instability in weakly magnetized disks. I Linear analysis, *Astrophysical Journal* 376 (1991), 214–22; John F. Hawley and Steven A. Balbus, A powerful local shear instability in weakly magnetized disks. II Nonlinear evolution, Astrophysical Journal 376 (1991), 223–33.
5. Steven A. Balbus, A classical thermal evaporation of clouds: An electrostatic analogy *Astrophysical Journal* 291(1) (1985), 518–22.
6. J. C. B. Papaloizou and J. E. Pringle, *Monthly Notices of the Royal Astronomical Society* 208 (1984), 721–50.
7. Steven A. Balbus, Dynamical, biological, and anthropic consequences of equal lunar and solar angular radii, *Proceedings of the Royal Society A* (2014), 470: 2168, 2014.0263.
8. H. M. Byrne, J. A. M. Green, S. A. Balbus, and P. E. Ahlberg, Tides: A key environmental driver of osteichthyan evolution and the fish-tetrapod evolution?, *Proceedings of the Royal Society A* (2020), 476: 2242, 2020.0355.
9. Richard Dawkins and Yan Wong, *The Ancestor's Tale*, 2nd edn, Weidenfeld & Nicolson (2016).

NOTES ON CONTRIBUTORS

Adam Jared Apt is a professional investment manager in Cambridge, Massachusetts, USA. He wrote his Oxford doctoral dissertation on the reception of Kepler's astronomy in England, and also supplied the Kepler entry (recently revised) in the Biographical Encyclopedia of Astronomers. He continues to publish on the history of science in early modern England, as well as on episodes in 19th-century American cartography and on aspects of investing. He is a past treasurer of the History of Science Society.

Natasha Bailey completed a DPhil degree in History at the University of Oxford in 2022 with a thesis titled 'Sleek Divines and Beaux Esprits: The Universities and the System of Knowledge in England, c.1690–1750', which she is revising for publication with Oxford University Press. She is currently working on a comparison of the activities of 'early career scholars' in the English and German speaking worlds as a postdoctoral researcher at the University of Erfurt. She has held a fellowship at the IZEA in Halle and has published in journals including *English Historical Review* and *History of Universities.*

Steven Balbus was Oxford's Savilian Professor of Astronomy from 2012 to 2024, retiring just before this volume went to press, and is currently a Senior Research Fellow at New College, Oxford. Widely recognized for his studies in astrophysical fluid dynamics, he received his undergraduate training at the Massachusetts Institute of Technology, reading mathematics and physics, and his PhD degree in theoretical astrophysics from the University of California at Berkeley. He is a Fellow of the Royal Society and a member of the US National Academy of Sciences. For his work on accretion disc turbulence, he received the 2013 Shaw Prize in Astronomy, jointly with his collaborator John Hawley, and for his contributions to theoretical astrophysics, he was awarded the Eddington Medal of the Royal Astronomical Society and the Dirac Medal from the Institute of Physics.

Sarah Dry is a writer and historian of science. She studied history and literature at Harvard College before completing an MSc degree at Imperial College and a PhD degree in the history of science at the University of Cambridge. Her previous books include *Waters of the World: The Story of the Scientists Who Unraveled the Mysteries of the Oceans, Atmosphere and Ice sheets – and Made the Planet Whole* (University of Chicago Press, 2019) and *The Newton Papers: The Strange and True Odyssey of Isaac Newton's Private Manuscripts* (OUP, 2013).

Pedro G. Ferreira is currently Head of Astrophysics at the University of Oxford, where he is also Professor of Astrophysics. His area of expertise is cosmology, in which he has pioneered investigations into the relic radiation left over from the Big Bang and on the nature of dark matter and dark energy in the Universe, and has pursued observational tests of alternatives to General Relativity theory. In addition to his research, Pedro writes extensively for general science audiences. His most recent book, *The*

Perfect Theory: A Century of Geniuses and the Battle over General Relativity, has been published in over twenty countries and was shortlisted for the 2014 Royal Winton Science Book Prize.

John Fisher came to academia in later life, a pioneer student of the Open University in 1971, gaining a first-class degree in the humanities from the Open University and an MA degree in Librarianship from City University, London. Following an MSc degree with distinction in the History and Philosophy of Science at Imperial College, he read for a doctorate with Rob Iliffe, and gained his PhD degree from Imperial College, London, in his 62nd year. He has recently published a book on the life and work of James Bradley.

Robert Goulding is an Associate Professor of Liberal Studies at the University of Notre Dame (Indiana, USA). He completed his PhD degree at the Warburg Institute, with a dissertation on Henry Savile's mathematical papers, part of which informed his 2010 monograph *Defending Hypatia: Ramus, Savile, and the Renaissance Rediscovery of Mathematical History*. He has published articles on Savile, Thomas Harriot's optics and mathematics, and the history of Platonism in late antiquity and the Renaissance. He is currently finishing a short book on Plotinus's natural philosophy.

Roger Hutchins is an alumnus of Magdalen College, Oxford. His DPhil thesis *British University Observatories 1772–1939* (published in 2008) explored extensively the experience of the Savilian professors over that time period, and made comparison to other universities in Britain and abroad. A Fellow of the Royal Astronomical Society and a Research Associate with, and contributor to, the *Oxford Dictionary of National Biography* (2008), he served for some years on the Council of the Society for the History of Astronomy.

Rob Iliffe is Professor of History of Science at Oxford University. He has published extensively on the history of relations between science and religion, the life and works of Isaac Newton, the history of scientific expeditions, changing conceptions of the nature and causes of scientific creativity, and the growth of the exact sciences between 1650 and 1850. He is co-director of the Newton Project, an online digital edition of Newton's writings.

Chris Lintott is a Professor of Astrophysics in the Department of Physics at the University of Oxford, and a Research Fellow at New College. His work involves the use of large astronomical surveys, especially through machine learning and citizen science, to study objects ranging from interstellar asteroids to galaxies. He is also the 39th Professor of Astronomy at Gresham College, London, and co-presenter of the BBC's long-running programme *The Sky at Night*.

Lee T. Macdonald holds a PhD degree in History and Philosophy of Science from the University of Leeds, and is the author of *Kew Observatory and the Evolution of Victorian Science 1840–1910*. He has recently completed a book about the history of the Royal Observatory at Greenwich in the period 1881–1939, the result of work that he carried out during a two-year research fellowship with Royal Museums Greenwich. He is currently Librarian and Archivist at the History of Science Museum, University of Oxford.

Robin Wilson is an Emeritus Professor of Pure Mathematics at the Open University, and of Geometry at Gresham College, London, and is a former Fellow of Keble College, Oxford University. A former President of the British Society for the History of Mathematics, he has written and edited over fifty books on mathematics and its history, including *Oxford's Savilian Professors of Geometry* and fourteen other books for Oxford University Press. Greatly involved with the popularization and communication of these subjects, he has received international awards from the Mathematical Association of America and other bodies for his 'outstanding expository writing' and his outreach activities.

PICTURE SOURCES AND ACKNOWLEDGEMENTS

Many people and institutions have helped in the preparation of this book, and in a variety of ways. In particular, the authors and editors are grateful to Oxford University's Bodleian Libraries for granting permission to reproduce manuscript images and copyright material, and also to the following Oxford University institutions for support, help, and advice: the Department of Astrophysics, the Mathematical Institute, the History of Science Museum, the Ashmolean Museum, and New College, Merton College, St Hugh's College, Green Templeton College, Keble College, and The Queen's College. We also thank the Royal Society, the Royal Astronomical Society, the International Seismological Centre, the Shaw Prize Organisation, the Starmus Foundation, the University of Aberdeen, Der Kongeline Bibliotek (Royal Danish Library), the Oxfordshire County Council Photographic Archive, Science Photo Library, DACS, and Bridgman Images.

We also wish to thank the following individuals: Max Alexander, Roger Bellamy, Phil Charles, Matthew Chipping, Danielle Czerkaszyn, Robin Darwall-Smith, Roger Davies, Amy Ebrey, George Efstathiou, Graham and Margaret Garnett, Jonn Godwin, Christopher Hollings, Jacqui Julier, Heather Kennedy, Robin Lane Fox, Tony Lynas-Gray, Stephanie Magnusson, Katherine Marshall, Alice B. Millea, Lance Miller, Thomas O'Rourke, William Poole, Pat Roche, Graeme Salmon, Joseph Silk, Michael Stansfield, Leona Stewart, Laura Søvsø Thomasen, Barrie Underwood, at Oxford University Press, Dan Taber and Giulia Lipparini, and at Integra Software Services, Nandhini Saravanan and Indumadhi Srinivasan.

Every effort has been made to trace all copyright holders, but if any have been inadvertently overlooked, the publishers will be pleased to make the necessary arrangements at the earliest opportunity,

FRONTISPIECE

[ii] Savile memorial: courtesy of The Warden and Fellows of Merton College, Oxford.

CHAPTER 1: SIR HENRY SAVILE AND THE EARLY PROFESSORS

[2] Bainbridge memorial: courtesy of The Warden and Fellows of Merton College, Oxford.

[4] Henry Savile: from R. T. Gunther, *Early Science in Oxford*, Vol. XI, Oxford University Press (1937), opposite p. 76.
Sir Henry Savile, engraving by Marcus Garrett: Wikimedia Commons.

[5] 'Cista mathematica': from R. T. Gunther, *Early Science in Oxford*, Vol. I, Oxford University Press (1923), opposite p. 122.

[6] Thomas Allen: R. T. Gunther, *Early Science in Oxford*, Vol. XI, Oxford University Press (1937), 286.

[7] Savile's *Prooemium mathematicum*: courtesy of the Bodleian Library, Savile Collection MS 29.

[8] Savile's lectures on astronomy: courtesy of the Bodleian Library, Savile Collection MS 31.

[10] Petrus Ramus: Wikimedia Commons.
Petrus Ramus, *Prooemium mathematicum* (Andreas Wechelus, 1567).

[13] Henry Savile, Lectures on Euclid (Oxford, 1621): courtesy of the Bodleian Library, Bod 1393 d.23.

[15] John Bainbridge: from R. Poole, *Catalogue of Portraits in the Possession of the University, Colleges, City, and County of Oxford*, Vol. II (Oxford, 1912), opposite p. 44.

[17] John Bainbridge and daughter, painting by Peter Lely, *c.*1640: Wikimedia Commons.

[18] Bainbridge's manuscript of Ulugh Beg: courtesy of the Bodleian Library, Savile Collection MS 46.
John Bainbridge *Canicularia* (Oxford, 1648): Wikimedia Commons.

[19] Henry Briggs, *Logarithmorum chilias prima* (Oxford, 1617).

[21] John Greaves: from R. T. Gunther, *Early Science in Oxford*, Vol. XI, Oxford University Press (1937), 48.
Diagram from *Pyramidographia* (London, 1646): Wikimedia Commons.

CHAPTER 2: INTERREGNUM AND RESTORATION

[22] Seth Ward: Wikimedia Commons.

[24] William Oughtred, portrait by Wenceslas Hollar: Wikimedia Commons.
William Oughtred, *Clavis mathematicae* (Oxford, 1667).

[26] John Wilkins: Wikimedia Commons.
John Wallis, portrait in the Examination Schools, University of Oxford: Wikimedia Commons.

[27] Wadham College: *The Oxford Almanac* (1738).

[29] Ismaël Boulliau: Wikimedia Commons.
Seth Ward, *In Ismaelis Bullialdi, Astronomiae philolaicae fundamenta* (Oxford, 1653).

[31] John Webster, *Academiarum examen, or the Examination of Academies* (London, 1654).
Seth Ward, *Vindiciae academiarum* (Oxford, 1654).

[34] Seth Ward as Bishop of Salisbury: Wikimedia Commons.

[35] Sir Christopher Wren: Wikimedia Commons.

[36] Solar eclipse: J. Wallis, *Eclipsis solaris Oxonii visae anno aerae Christianae 1654* (Oxford, 1655), 2.

[39] Wren's engine for grinding lenses: *Philosophical Transactions* 4 (no 53) (15 November 1669), facing page 1059.

[40] Wren's drawing of the brain: from Thomas Willis, *Cerebri Anatome* (London, 1664).

[41] Wren's sundial at All Souls College: Wikimedia Commons.

[42] Sheldonian Theatre, engraving by David Loggan, 1675: Wikimedia Commons.

Sheldonian Theatre interior, engraving by J. Skelton, 1820: Wikimedia Commons.
[44] Edmond Halley, *Apollonii Pergaei conicorum libri octo* (Oxford, 1710).

CHAPTER 3: THE NEWTONIANS

[46] Memorial to David Gregory, photograph by John Fauvel.
[49] Gregory inaugural lecture: courtesy of the University of Aberdeen Collections (under CC-BY4.0), MS 2206/3/3, f.27r.
[52] David Gregory, *Astronomia physicae & geometricae elementa* (Geneva, 1726).
[53] David Gregory, *Catoptricae et dioptricae sphaerica elementa* (Oxford, 1695).
Dr. Gregory's Elements of Catoptrics and Dioptrics (London 1715).
[55] Gregory's edition of the works of Euclid (Oxford, 1703).
[56] Handwritten note from Gregory's papers: Wikimedia Commons.
[57] John Caswell, *Account of the Doctrine of Trigonometry* (Oxford 1685).
[59] Thomas Burnet, *Telluris theoria sacra*: Wikimedia Commons.
John Keill, *An Examination of Dr. Burnet's Theory of the Earth* (Oxford, 1698).
[60] John Keill's *Introductio ad veram physicam* (Oxford, 1702).
[63] John Keill, *Introductio ad veram astronomiam* (Oxford, 1718).
[65] James Keill, *An Account of Animal Secretion* (London, 1708).

CHAPTER 4: JAMES BRADLEY

[68] James Bradley. mezzotint by J. Faber: Wikimedia Commons.
[70] Bradley's *Works and Correspondence*, edited by S. P. Rigaud (Oxford, 1832): Wikimedia Commons.
[72] Wanstead House: Wikimedia Commons.
[75] Engraving of George Graham: Wikimedia Commons.
[78] House in New College Lane: from R. T. Gunther, *Early Science in Oxford*, Vol. II, Oxford University Press (1923), 83.
[79] Course of lectures: from R. T. Gunther, *Early Science in Oxford*, Vol. XI, Oxford University Press (1937), 404.
[80] Shirburn Castle, engraving by J. Neale (1818): Wikimedia Commons.
[84] Bradley's New Observatory: from J. Bernoulli, *Lettres astronomiques* (Berlin, 1771).
[85] John Bird, mezzotint by V. Green, 1776: Wikimedia Commons.
Bird's 8-foot quadrant: Wikimedia Commons.
[86] Nathaniel Bliss portrait: Wikimedia Commons.
Drawing of Bliss, from R. T. Gunther, *Early Science in Oxford*, Vol. XI, Oxford University Press (1937), 276: Wikimedia Commons.
[87] Longitude table from *The Nautical Almanac and Astronomical Ephemeris for the Year 1767*: Wikimedia Commons.
[88] Friedrich Wilhelm Bessel engraving, 1867: Wikimedia Commons.

CHAPTER 5: ENLIGHTENMENT PROFESSORS

[90] Interior of Radcliffe Observatory: Wikimedia Commons.
[93] Portrait of Thomas Hornsby: reproduced with the permission of the Principal and Fellows of Green Templeton College, University of Oxford.
[95] Advertisement for Hornsby's lectures: courtesy of the History of Science Museum, University of Oxford.

Hornsby's lecture on the transit of Venus: from R. T. Gunther, *Early Science in Oxford*, Vol. XI, Oxford University Press (1937), 407.

[96] Hornsby's quadrant and transit: from R. T. Gunther, *Early Science in Oxford*, Vol. II, Oxford University Press (1923), opposite p. 318.

[97] John Radcliffe engraving: Wikimedia Commons.

John Radcliffe, statue by Martin Jennings: photograph by Christopher D. Hollings.

[98] John Dollond portrait: Wikimedia Commons.

[99] Radcliffe Observatory, courtesy of Robin Wilson.

Atlas at the Radcliffe Observatory: Wikimedia Commons.

[100] Thomas Bugge's sketch of the Observatory and meridian marker: courtesy of the Royal Danish Library, NKS 377e folio, fol. 58r & 60v.

[101] Bugge's sketch of a transit room: courtesy of the Royal Danish Library, NKS 377e folio, fol. 60v.

[102] Hornsby's Sedleian lectures: from R. T. Gunther, *Early Science in Oxford*, Vol. XI, Oxford University Press (1937), 410.

[103] Bird's zenith sector and observing chair: R. T. Gunther, *Early Science in Oxford*, Vol. II, Oxford University Press (1923), opposite p. 326.

[106] A. Robertson, *Elements of Conic Sections* (Oxford, 1818).

Robert Peel, from Anne E. Keeling, *Great Britain and Her Queen*, London (1887), Ch. II.

[110] Stephen Peter and Mary Rigaud, painting in the Ashmolean Museum: Wikimedia Commons.

Silhouette of Rigaud, from R. Poole, *Catalogue of Portraits in the Possession of the University, Colleges, City, and County of Oxford*, Vol. II (Oxford, 1912), opposite p. 44.

[111] Meridian circle: from R. T. Gunther, *Early Science in Oxford*, Vol. II, Oxford University Press (1923), opposite p. 326.

[112] Stephen Peter Rigaud, *On the Arenarius of Archimedes*, Oxford (1837), title page.

CHAPTER 6: THE VICTORIANS

[114] Charles Pritchard: frontispiece from Ada Pritchard (ed.), *Charles Pritchard, Late Savilian Professor of Astronomy in the University of Oxford. Memoirs of his Life* (London, 1897).

[117] The new University Museum: *Oxford Almanac*, 1860.

[120] Appointments of George and Manuel Johnson: *Oxford City and County Chronicle* (4 May 1839).

[122] Conferral of the Deanery of Wells to Johnson: *Buckingham Herald* (11 March 1854).

[123] George Johnson, Dean of Wells: courtesy of The Provost and Fellows of the Queen's College, Oxford.

[124] William Donkin around 1843: courtesy of the Bodleian Library, MS Photog. B34 ff 133.

[125] William Donkin's election in May 1842: courtesy of the Bodleian Library, OUA/NEP/A/1.

[126] The Radcliffe heliometer, drawn by Annarella Smyth: from Vice-Admiral W. H. Smyth, *Speculum Hartwellianum*, 1860.

[127] Donkin's stipend in 1858: courtesy of the Bodleian Library, OUA UC6/38.

[128] University Museum Keeper's house, courtesy of the Museum of Natural History, University of Oxford.
Museum Observatory: courtesy of the History of Science Museum, University of Oxford, MS Museum 117.
[129] Oxford's University Museum interior: *Illustrated London News* (6 October 1860).
[130] William Donkin around 1868: Wikimedia Commons.
[131] Donkin's death announcement: *The Oxford Times*, 20 November 1869.
[135] Charles Pritchard: from R. T. Gunther, *Early Science in Oxford*, Vol. XI, Oxford University Press (1937), 142.
[136] Warren De la Rue: Wikimedia Commons.
[137] New Savilian Observatory: courtesy of the Oxfordshire County Council, Oxford History Archive, POX 0088792.
[139] 'A Star in the East': courtesy of the Royal Astronomical Society/Science Photo Library.
[141] Oxford University's Grubb refractor: courtesy of the Oxfordshire County Council, Oxford History Archive, POX 0116598.

CHAPTER 7: THE 20TH CENTURY

[144] Herbert Hall Turner: courtesy of the Bodleian Library, MS Photog. 1c60, f.35, Walter Bennington Estate.
[147] Paris Astrographic congress of 1887: Wikimedia Commons.
[149] Frank Bellamy with the De la Rue reflector: courtesy of the Bellamy family and the History of Science Museum, University of Oxford.
[151] H. H. Turner, © Estate of Gilbert Spencer. All rights reserved, DACS 2025.
Herbert Hall Turner: from R. T. Gunther, *Early Science in Oxford*, Vol. XI, Oxford University Press (1937), opposite p. 145.
[153] Turner with seismographic globe: courtesy of the International Seismological Centre.
[153] Ethel Bellamy receiving her MA degree: courtesy of the Bellamy family and the History of Science Museum, University of Oxford.
[154] Herbert Turner memorial plaque: photograph by Steven Balbus.
[155] Harry Hemley Plaskett: courtesy of Lafayette photography.
[158] Madge Adam: by kind permission of the Principal and Fellows of St Hugh's College, Oxford, SHG/M/2/5/1948.
[159] New College engraving: 10271/8 © courtesy of the Warden and Fellows of New College Oxford, NCA 5597.
[162] D. E. Blackwell photograph: courtesy of the Royal Astronomical Society/Science Photo Library.
[164] D. E. Blackwell, 1982 (pencil on paper), Winer, Marc S. A. (1943–85)/courtesy of the Warden and scholars of New College, Oxford/Bridgeman Images.
[165] Warden and Fellows of New College, 1981: by permission of Gillman & Soame photographers and the Warden and Fellows of New College Oxford, NCA SCR/A3/7.
[166] Denys Wilkinson building: Wikimedia Commons.

CHAPTER 8: THE GOLDEN AGE OF COSMOLOGY

[168] Joe Silk on the cover of *Scout* magazine: courtesy of *The Scout*.
[171] Arthur Milne, Roger Penrose, and Dennis Sciama: Wikimedia Commons.
[172] Joseph Silk: courtesy of Joe Silk.

[173] Joseph Silk paper: *Astrophysical Journal* 151 (1968), 459–71.
[174] Joe Silk at Woods Hole: courtesy of Joe Silk.
[176] 3-year-old George Efstathiou with parents: courtesy of George Efstathiou.
[177] George Efstathiou and Dick Yarrow: courtesy of George Efstathiou.
[178] George Efstathiou and Richard Ellis: courtesy of George Efstathiou.
[179] 'Gang of Four': courtesy of George Efstathiou.
[182] 2df Galaxy Redshift Survey: Wikimedia Commons.
[183] George Efstathiou at Savilian leaving party: courtesy of George Efstathiou.
[187] George Efstathiou with Planck Science team: courtesy of George Efstathiou.
[188] Joseph Silk: courtesy of Joe Silk.
Joseph Silk's *Back to the Moon*: Princeton University Press, 2022.
[189] Joseph Silk receiving the RAS Gold Medal: courtesy of the Royal Astronomical Society.

CHAPTER 9: INTERVIEW WITH STEVEN BALBUS

[190] Steven Balbus: © The Royal Society.
[192] Steven Balbus with book: courtesy of Steven Balbus.
Steven Balbus ice skating: courtesy of Steven Balbus.
Steven Balbus's 21st birthday: courtesy of Steven Balbus.
Steven Balbus at MIT: courtesy of Steven Balbus.
[194] Steven Balbus receiving the Shaw prize: courtesy of the Shaw Prize Organisation.
[197] Steven Balbus with Caroline Terquem: courtesy of Steven Balbus.
[198] Steven Balbus with Martin Rees and Andrew Fabian: © The Royal Society.
[199] Steven Balbus and Richard Dawkins: courtesy of the Starmus Foundation and Max Alexander.

INDEX

U

V

W

Y

Z